Crop Protection Handbook – Cereals

Crop Protection Handbook – Cereals

P. J. ATTWOOD

BCPC Publications
2A KIDDERMINSTER ROAD, CROYDON CR0 2UE, ENGLAND

British Library Cataloguing in Publication Data
Crop protection handbook – cereals.
1. Plants, Protection of
I. Attwood, Peter II. British Crop Protection Council
632 SB950

ISBN 0-901436-97-6

Printed in Great Britain by
The Lavenham Press Limited
Lavenham, Suffolk

CONTENTS

CONTRIBUTORS

D. H. K. Davies
The East of Scotland College of Agriculture

R. GAIR
Formerly ADAS Regional Entomologist, Cambridge

R. H. JARVIS
Director, Boxworth Experimental Husbandry Farm, ADAS

B. J. WILSON
Long Ashton Research Station, Weed Research Division

D. J. YARHAM
ADAS Regional Plant Pathologist, Cambridge

FOREWORD

The British Crop Protection Council and its two parent Councils have been publishing independent advice and information on weeds, pests and diseases and their control for more than 30 years, a period covering almost the whole of the remarkable development of agrochemicals in modern agriculture. This book thus has a long pedigree and its contributors have been able to draw on an unrivalled wealth of accumulated expertise and knowledge.

The story began in 1953 when the British Weed Control Council was formed following publication of the first truly independent and authoritative recommendations for the use of the few weedkillers then available, recommendations developed by the pooling of experience by government and university, research workers, manufacturers and distributors of pesticides, agricultural contractors, advisers and users. From this small beginning grew the well known *Weed Control Handbook* and later the *Insecticide and Fungicide Handbook*, volumes used and respected throughout the world for the comprehensive, authoritative, up-to-date and, perhaps above all, independent information and guidance they provided on crop protection. Each book covered all major crops grown in the UK, as well as many minor ones, and gave particular emphasis to chemical methods of control.

In recent years, the technology of crop protection has expanded so much that it is no longer feasible, nor even helpful, to contain in a single volume the essential information on weeds, pests and diseases and on the multitude of control agents relevant to all the numerous crops grown in this country. Moreover, as farming and horticultural systems of crop production become more specialised the need to consider crop protection in the round, as distinct from controlling each weed, pest or disease organism individually on an *ad hoc* basis, increases. The grower and his adviser require the information they need to formulate a rational crop protection strategy for the specific cropping system in question to be available from a single source. They should not have to search for it in several different and possibly conflicting publications. The present-day economic pressures on farming and the growing trend to combine crop production with conservation of landscape and wild-life also demand a more informed approach to crop protection than hitherto, and a better understanding of the interactions between weeds, pests and diseases, the husbandry systems employed and the application of agrochemicals: all within the context of individual farming systems.

Thus, for many reasons the British Crop Protection Council decided to introduce a new series of Handbooks, each providing a comprehensive treatment of the weeds, pests and diseases of a particular group of crops and of the options available to the grower for controlling, or avoiding them, with enough information and discussion on principles to allow extrapolation to his particular circumstance.

This is the second Handbook in the series. As in the case of the earlier Handbooks of the Council, it is very much a joint effort by leading experts from the research and advisory fields and from the agrochemical and farming industries. It goes without saying that as a BCPC publication the information it contains is unprejudiced, independent and highly authoritative.

As cereal growing comes under increasing strain, and profitable production proves more difficult to sustain, feel sure this book will prove of enormous value to all growers anxious to stay in the business, to their advisers and to all whose job it is to provide and to purvey the flow of information and technology on which British agriculture depends.

J. D. FRYER
Sandacre, Woodside
Abingdon, Oxon

EDITOR'S PREFACE

Ever closer attention needs to be given to the systems employed in growing cereals to ensure that they are sustainable at economic costs of production. In considering changes to systems, many inter-relating factors have to be borne in mind not least of which are the effects of weeds, pests and diseases.

Previous British Crop Protection Handbooks, the *Weed Control Handbook* and the *Pest and Disease Control Handbook*, have dealt mainly with the principles and practice of using crop protection chemicals. This Handbook breaks with that tradition by including more about husbandry or cultural control methods and their integration with chemical use. As is pointed out in Chapter 1 there is nothing very new about integrated control methods, but we do seem to be entering into a period when understanding and using them is becoming increasingly important to meet economic and environmental needs.

This book is organised in a way which should enable the reader to see how weed, pest and disease control methods inter-relate and how husbandry factors differ in their effects on cereal crop problems. The introductory chapter puts integrated weed, pest and disease control in perspective, while Chapter 2 describes the main problems encountered in cereals. The third chapter discusses husbandry as an aid to integrated control; this is followed by chapters introducing the topic of chemical crop protection and the application of the materials.

The final chapters contain outline recommendations for control measures arranged in a way which makes it possible to plan ahead to the next critical period when several operations may need carrying out in the same period. It is hoped that the tabulated guides to crop protection planning and crop scouting will be a useful *aide mémoire* to items which should be considered at different stages of winter and spring cereal growth.

The organisation of the recommendations and the guides to crop protection planning were one of the many bright ideas of the late J. G. Elliott. His help is acknowledged with thanks along with the authors named elsewhere who were assisted by many people including the ADAS Plant Pathology Group, Dr H. Allen, S. A. Evans, Dr C. G. L. Furmidge, Dr.V. W. L. Jordan, L. E. W. Stone, W. A. Taylor and E. W. Woolley. Finally my thanks are due to the patient encouragement and help of the BCPC Managing Editor, Dr E. K. Woodford, and his wife.

ERRATA SLIP

Crop Protection Handbook—Cereals

Editor's Preface, p.xii :

The list of helpers should include Dr. R. G. McKINLAY

NOTE TO READERS

The common names of plants used in this book are those given by Dony, Rob and Perring in *English Names of Wild Flowers* (Butterworths, London). The Latin names of plants also follow the same reference in the main. Some inconsistencies arise due to the retention of a generic name when it is commonly understood as representing a group of weeds in cereals.

The recommendations for crop protection chemicals are purposely brief and short of detail, not only to keep the book within a reasonable size, but also because the product label is the definitive statement which should be followed. Appendix III provides a means of translating chemical names into products.

The growth stage numbers used in this book are based on the system devised by Zodaks, Chang and Konzak, published in 1974 as "A decimal code for the growth stages of cereals" (*Weed Research* **14,** 415–421). This was further developed by Tottman and Makepeace in 1979 as "An explanation of the decimal code for the growth stages of cereals with illustrations" published in *Annals of Applied Biology* (1979) **93,** 221–234, which was reprinted as Occasional Publication No. 2 by BCPC Publications, Croydon.

LIST OF TABLES

Chapter 1 INTEGRATED CONTROL

The need to control pests, diseases and weeds in cereals has long been recognised. Early systems of agriculture used techniques such as shifting cultivation to achieve this end as well as to maintain adequate plant nutrient levels in the soil. Following centuries of development, pests, diseases and weeds in cereal crops can now be controlled by agrochemicals or by a range of cultural methods such as crop rotation and cultivation. The different methods may interact with one another and should be brought together to form an integrated control programme. Integrated control involves selecting the techniques best suited to a particular cropping system and incorporating them into an overall programme which exploits the strengths and compensates for the weaknesses of its different components. It should avoid undue reliance on control methods which may have adverse effects on wild-life and environmental health. Any aspect of the husbandry of the crop which may affect the incidence of pests, diseases and weeds, even if it is not a specific part of the control programme, must be considered when planning the overall strategy.

Integrated control is not a new concept. Both the open field agriculture of the mediaeval period and the "high farming" systems of the late 18th and 19th centuries depended on the integration of a variety of husbandry techniques, particularly rotation and cultivation, to maintain all aspects of fertility including freedom from pests, diseases and weeds. Some of the cultural practices which farmers once relied on, such as inter-row cultivation in cereals, are no longer used because of changed economic and technological circumstances. Others are still needed to deal with pests, diseases and weeds which cannot be controlled by chemical sprays or because they offer cheaper or more effective alternatives in a particular farming situation. Far from reducing the need for an integrated approach, the introduction of modern agrochemicals has increased the number of options and so added to the complexity of planning an integrated programme.

The principles of integrated control can be illustrated by an example confined for simplicity to weed control. On heavy land, a farmer may be growing cereals in rotation with oilseed rape and beans; all the crops are sown in autumn. To counter the problem of completing seedbed preparation and sowing in the short time available, he relies on shallow cultivation rather than ploughing. Because he has chosen not to include spring-sown crops in the rotation and not to plough, the main methods of weed control available to him are confined to rotation and herbicides. With no difference in sowing season, the principal value of crop rotation lies in the change of herbicide type which is possible in the non-cereal breaks. It is on these that he must rely for removing annual grass weeds which cannot be controlled selectively in the cereals or in the short interval between harvest and autumn sowing. In general, all grasses are easier and cheaper to control chemically in non-cereal crops than in cereals, while the latter offer better opportunities for controlling broad-leaved weeds. The herbicide programme should be planned to take account of these points. In addition, the farmer must be prepared to use herbicides more skilfully and perhaps more frequently than if he grew a wider range of crops. He must also realise that reduced cultivation may lower the efficiency of soil-acting herbicides and check that the varieties of cereals which he selects are not susceptible to damage from the herbicides he intends to use. In practice, it is of course necessary to make similar plans for pests and diseases as well as weeds and also to consider interactions between the three.

Principles of integrated control

In any integrated system of pest, disease and weed management there are several general principles which apply. The primary aim of a farm business is to maintain and preferably improve the profit margin both in the short and long term and the control of pests, diseases and weeds is one of the many aspects which contribute to this overall target. At the most basic level, it follows that where more than one control measure exists the cheapest will be preferred but the choice is rarely so simple. For instance, alternative methods usually differ in their effectiveness; a serious problem will require the most effective solution even if it is expensive, while a minor problem may be contained by relatively cheap measures which give a poorer degree of control. Secondly, a farmer's choice is usually influenced considerably by the reliability of different methods as well as by their cost. The uncertainty due to the inherent dependence of cereal production on weather conditions leads to a reluctance to take additional risks and frequently a more expensive alternative is preferred because it is likely to be more certain in its results. Thirdly, most methods of pest, disease and weed control have more or less important implications for other aspects of the husbandry of the crop or for the farming system in general. The most obvious example is rotational control which when practised in the form of the Norfolk four-course and other similar rotations imposed serious restraints on the whole farm business and allowed little or no flexibility to meet changing economic conditions. On a very different level of importance, a decision to use a particular spray chemical may restrict choice of variety and of other agrochemicals to be used in the same spray programme.

In order to improve the overall profitability of the enterprise, husbandry methods which may themselves increase pest, disease or weed problems may be used. In the example cited above of a farmer on heavy land growing only autumn-sown crops and relying on reduced cultivations, the options of growing spring-sown crops and ploughing regularly have been foregone in the interest of improved profitability. A farmer cannot be criticised for choosing husbandry methods which will add to the success of his business provided that he has considered fully the implications of his chosen system in terms of pest, disease and weed control. He must, however, always be prepared to make changes to meet changing conditions. Modifications to the pattern of cropping or cultivations methods, for instance, may affect the optimum balance between cultural and chemical control measures and so make it necessary to restructure the integrated programme.

Objectives of control programmes

Within the overall objective of maximum profitability a farmer should have detailed objectives in mind when planning a pest, disease and weed control programme. In broad terms the aim should be prevention, eradication or containment at levels where the value of any yield loss will be less than the cost of control. However, weed control may also be justified because it increases the rate of combine harvesting rather than giving rise to a measurable yield increase. In cereals the scope for prevention or marked reduction in the incidence of a pest, disease or weed is limited by practical problems and cost, although there are some exceptions. For instance, the risk of barley yellow dwarf virus in winter cereals can be minimised by avoiding early sowing; cereal leaf diseases can be reduced considerably by choosing resistant varieties. On the other hand, full preventive spraying against cereal pests, diseases and weeds is unlikely to be cost effective. Preventive or insurance spraying is also undesirable on environmental grounds and may encourage the development of resistant strains of pest and disease organisms.

There are few opportunities for eradicating pests, diseases and weeds in cereals although some farmers have made good progress towards ridding their holding of wild-oats. The cost of eradication may reduce profit in the short term but improve it in the longer view. Eradication should be considered only when it can be followed by effective measures to prevent re-introduction.

Threshold levels

In most situations containment offers the best policy for pest, disease and weed control. In several cases threshold levels, that is levels at which the cost of control is likely to be recouped by the increase in yield, have been established. Examples include pests such as grain aphids, diseases such as eyespot and weeds such as wild-oats. The threshold may not be concerned only with the number of organisms present; in the case of the aphid vector of barley yellow dwarf virus, the infectivity of the aphids is also taken into consideration. Much work remains to be done in establishing threshold levels for other pests, diseases and weeds. In many cases it is difficult to predict, at the time when an organism can be controlled, the extent to which it will develop and affect yield. For instance, the competitive effect of a given population of black-grass seedlings varies with the density and vigour of the crop; the effect of eyespot on yield depends on weather conditions during the whole of the growing season. The "control threshold" must be lower than the economic threshold to allow for these variations. This aspect is discussed more fully, as regards weeds, in Chapter 2 page 10.

Even when threshold levels have been satisfactorily established, a flexible approach is required in planning control measures. Problems are least with weeds and soil-inhabiting pest and disease organisms; their probable incidence can be assessed by inspecting previous crops and provision made in advance for containment. With winged insects and fungi spread aerially or by rain splash, much less warning of the need for control measures will be available. Nevertheless, the establishment of threshold levels for a wider range of pests, diseases and weeds is essential to achieve effective control without the expense and possible hazards of insurance spraying.

It must of course be acknowledged that crop protection is not an exact science. In addition to the variables noted above which affect the incidence and development of pest and disease attacks and weed problems, the effectiveness of control measures is influenced by current weather and soil conditions. In the extreme, either of these may prevent the application of a planned treatment. Any containment policy must therefore include a reasonable margin to allow for the unexpected failure of at least part of the planned control programme.

The detailed form of containment measures may vary considerably. If the number of possible methods of dealing with a pest, disease or weed is very limited and particularly if control will be expensive or interfere with other husbandry operations, a short intensive control programme to reduce the problem to negligible levels may allow a respite of several years before further treatment is needed. The control of couch grass in an intensive cereal system usually follows this pattern. In other cases a routine annual treatment may be preferred to contain the pest, disease or weed at insignificant levels.

Level of control required

Other important factors in deciding the form and extent of a containment programme are the importance of the commodity in the farming system and the potential yield of an individual crop. On a farm where cereals predominate, failure to achieve satisfactory pest, disease and weed control in them will jeopardise the whole farm income. In a more traditional system which includes a wider range of crops grown in rotation and probably at least one livestock enterprise, a reduction in cereal yields, although unwelcome, is less serious because of income from other components of the system. Similar considerations apply to cereal crops with a particularly high yield potential grown in any type of farming system. A farmer who has lavished management skill and high level of inputs such as fertilizer on a field may reasonably hope for a yield 60-70 % higher than from the average crop. He will be more inclined to be generous with pest, disease and weed control measures and will choose methods which have a lower risk of failure than alternative techniques even if they are more costly.

The more generous use of pesticides, fungicides and herbicides in circumstances such as these is not unreasonable but in some cases the concept of insuring the investment already made in the crop by applying more and more sprays has been taken too far. Instead of containment, prevention has become the aim. The dangers of this approach have already been stressed. It is particularly undesirable if prevention depends almost entirely on chemical control. Although the opportunities for husbandry methods of control may be more limited than in the past some, such as rotation and sowing date, are very potent and it is particularly desirable to utilise them. An integrated programme including both chemical and cultural techniques is essential if the most cost effective control of pests, diseases and weeds is to be achieved. Further, full cost-effectiveness will be maintained only if the balance between chemical and cultural control is adjusted when necessary to suit changing price structures within the industry.

Chapter 2 THE SIGNIFICANCE OF WEEDS, DISEASES AND PESTS IN CEREALS

Weed Problems

How weed problems arise

Weeds which emerge in a cereal crop are a historical legacy from previous years. They arise from seeds, stems, roots or bulbs in the soil which in turn were produced by weeds which survived control measures in previous crops. The weed flora which evolves is greatly influenced by the system of cropping and cultivation. A wide diversity of species may be expected where frequent changes in the rotation have been practised in the past. In contrast, monoculture with an unchanged cropping and cultural system will encourage relatively well adapted species to build up to high numbers and to become dominant. The recent trend towards autumn-sown cereals for example has favoured those weed species which naturally germinate in the autumn. Reduced cultivations have favoured those species which are prolific seeders, which produce seeds of relatively short dormancy, and which germinate and emerge readily from the surface layers of the soil but not from depth. Seeds of some of these species are vulnerable to straw burning but this practice alone will not usually stop populations increasing. This can only be prevented by the effective use of herbicides, and any weed species not readily controlled is likely to become a major problem.

Effect of cultivations on the weed flora

The cultivation system can have a profound effect on the weed flora. Continuous direct drilling will tend to result in fewer broad-leaved weeds than where the land is ploughed, with tine cultivation having an intermediate effect, although this trend may not apply to common chickweed and the mayweeds. The influence of tillage on grass weeds is much more pronounced, and black-grass, wild-oats and brome are all significantly encouraged by reduced cultivations compared with ploughing. Species which produce a lot of seed which is relatively non-dormant are able to exploit the situation of having seeds retained in the topsoil from where they can readily establish new plants.

Stubble cultivations immediately after harvest have in the past been an important feature of systems for the control of rhizomatous couch grasses. Such cultivations can stimulate seeds of annual weeds to germinate and produce plants which are killed by subsequent operations. However, burial by cultivations induces seed dormancy in many species, and although a small proportion of seeds may be encouraged to germinate the majority will survive as buried dormant seeds. Stubble cultivation has been shown to favour the build up of wild-oats. Stubbles containing large numbers of freshly shed seeds are best left uncultivated during the autumn to encourage natural losses of seeds. The exception is where brome grass is a problem since germination of brome seed can be delayed by light. With little or no innate dormancy, it is better to incorporate the seed into a fine seedbed to encourage rapid germination, so that the plants are killed by subsequent operations.

Biological reasons for the success of cereal weeds

The main weeds likely to be encountered may be divided into annual and perennial broad-leaved and annual and perennial grass weeds. Within each group it is important to recognise the main biological features which contribute to their success as weeds. The main reasons for success are the survival and competitive ability within the crop and the characteristics of the seeds or vegetative organs by which the weeds reproduce.

Those annual weeds which emerge in the early stages of crop growth are the most important. They will be larger, more competitive and produce more seeds than those emerging when the crop is well established. In well established winter cereals, spring germinating weeds are less likely to reduce crop yields but may still be significant in terms of seed return. Important features of successful annual weeds are the early shedding of seeds before cereal harvest, the self burial of seeds on the soil surface, e.g. wild-oats, and seed dormancy to ensure a reserve of seeds in the soil predisposed to germinate only in favourable conditions. Seeds of broad-leaved species generally persist longer in the soil than do seeds of grass species.

Perennial weeds generally emerge later than annual weeds, when the crop is well established. They can become a significant problem during the late spring and summer, making vigorous growth when the crop canopy breaks especially if moisture is not limiting.

Significance of weeds

Competition. Cereals are most sensitive to competition from weeds in the early stages of crop growth, and particular note should be taken of annual weeds emerging with the crop. As a general principle weeds which emerge with the crop will be more competitive than those emerging once the crop is established. In early-drilled winter cereals such weeds will be better established, more likely to overwinter and be larger and more difficult to kill in the spring than weeds in late-drilled crops.

Weeds compete with the crop for light, space, moisture and nutrients. Early competition from large numbers of weeds is mostly for nutrients. Competition for moisture and light become important later when weeds have produced abundant foliage. Late emerging weeds which remain below the canopy of a vigorous crop are unlikely to reduce crop yields.

Competition is affected by the balance of factors which influence the relative growth of crop and weeds. Dry weather after sowing the crop can reduce the number of weeds which germinate near the surface and allow the crop to pass through its sensitive early stages relatively free from weed competition. However, germination of the cereal may be delayed if sown into very dry soil; weeds emerging from deeper layers can grow unchecked and later compete strongly. Mild winters favour autumn germinating weeds such as common chickweed and cleavers which can maintain slow growth throughout the winter to become very competitive in the early spring. Weeds become better established and are more likely to overwinter in crops sown early rather than late in the autumn. Crops sown too deeply are slow to emerge, and this may give shallow-germinating weeds a competitive advantage. Competition may be greater in short-strawed cereal varieties where weeds can receive more light through a more open cereal canopy. Vigorous cereal crops suppress weed growth well. Black-grass and wild-oats in particular respond to low crop vigour by greater tillering, vegetative growth and seed production. Establishing a good population will help to counter the effects of early weed competition. This is important where dense weed infestations are expected; low seed rates or poor crop establishment can help to maintain the problem. Average or above average rates of seeding should be used to guarantee a crop population of at least 250 seedlings/m^2.

Effect on harvesting. The grain from a crop infested with weeds costs more to harvest, clean and dry than grain from a clean crop. Weedy crops take longer to dry out, and slow the combine; this is particularly serious in a wet season. Early weed competition can weaken the crop and pre-dispose it to lodging. Climbing weeds such as cleavers and bindweed and grass weeds such as black-grass and brome when present in abundance can cause lodging which can lead to late weed growth in the lodged areas creating great problems at harvest. Grain from such crops is likely to be of a higher moisture content, particularly if the weeds at harvest are vegetative and immature.

Weeds as grain contaminants. The seeds of some weeds such as wild-oats and cleavers are difficult and costly to remove from harvested grain. The level of contamination of wild-oat seeds is likely to be higher when crops are harvested early before many seeds have shed. Freedom from weeds is particularly important in crops grown for seed, and regulations limit the amount of contamination allowed in seed offered for sale. Home-saved seed should also be produced to a high standard of cleanliness because a relatively low percentage by weight of weed seeds in the seed grain can result in a large number of weeds being sown.

Weeds as carriers of diseases and pests. Important connections between weeds, diseases and pests are discussed in later sections of this chapter. Weed grasses are more likely to be troublesome to cereal crops in this respect than broad-leaved weeds. There is little firm information on the relative importance of weeds as carriers of pests and diseases compared with other agents.

Threshold weed populations

Even low weed populations will affect a crop to some degree. The *economic threshold* is the weed population where the cost of spraying is recovered by the extra yield obtained as a result of spraying. This will be largely determined by the relationship between the price received for grain and the cost of herbicide treatment, but other economic advantages of using herbicides such as ease of harvesting and obtaining grain free from weed seeds may lower the threshold. This concept assumes that at a population just below the threshold it would not be economically worthwhile to spray, and yet the farmer will still tolerate some financial loss due to weed competition. For example he may always be willing to spend say £30/ha to control a threshold population of 10 wild-oats/m^2, but he would tolerate 9/m^2 and the consequent yield and financial loss. In practice this is rarely true and the economic threshold is very much a theoretical concept.

Precise definition of an economic weed threshold is difficult because of the variability in competitive effects due to the many factors outlined in the preceding section. These cannot all be predicted and systems of management must be flexible to accommodate variations in competitiveness and herbicide performance. In practice, the farmer will operate a "spray decision rule" at a lower population than the economic threshold. This decision must take into account the potential increase of the weed if not controlled under the particular management system. This spray decision rule or "farmer's threshold" will be subjectively based on the farmer's own experience. This approach to spraying decisions is likely to become increasingly important if cereal profitability declines.

Spray decisions based on thresholds are dependent on a system of whole field population monitoring. With grass weeds it is difficult to recognise low populations of seedlings in a cereal crop in order to make immediate decisions on whether or not to spray. It is easier to count grass seed heads in June or July so that spraying decisions can be made in advance for the next crop. This forecast or "predictive decision rule" assumes a relationship between the seed heads present and the numbers of seedlings emerging in the next crop.

Threshold levels for the major grass weeds are discussed in later sections; little information is available for broad-leaved species. Threshold levels for spraying mixed weed populations have not been defined, but recent work at the Weed Research Organisation suggests it might be possible to integrate the competitive abilities of several species in a mixed infestation. This is based upon a system of crop equivalents whereby the density of each species is adjusted for its "competitive index", (i.e. the extent to which a weed of that species can replace a cereal plant), and is derived from the dry weight of the weed relative to the crop. The total mixed weed population can then be expressed as a single crop equivalent value. This system will need further development over a number of years.

Weed occurrence and distribution

An indication of the occurrence and regional distribution of a number of cereal weeds in the UK is given in The British Crop Protection Council's (BCPC) *Weed Control Handbook: Principles*, 1982 7th Edn, Blackwell, Oxford. Although a wide number of species occur in cereals, only relatively few become dominant. Wild-oats, cleavers, mayweeds and common chickweed are the most widespread. Meadow and couch grasses are the most important grass species in the west and north of England and in Scotland; wild-oats and black-grass in the east of England. Common hemp-nettle is a major broad-leaved weed problem in spring-sown cereals in the north and in Scotland, and cleavers in the south and east of England. Knotgrass, redshank, pale persicaria, black-bindweed and common chickweed are frequent in spring crops in all areas.

Main annual weeds of winter cereals

The main annual weeds likely to be encountered at the time of emergence of winter cereals are the annual grasses such as wild-oats, black-grass, meadow grasses and the brome grasses, together with an association of broad-leaved species well adapted to autumn-sown cereals whose germination periods coincide with autumn sowing. The most common are cleavers, common chickweed, field pansy, field forget-me-not, common poppy, wild radish and the mayweeds, deadnettles and speedwells.

Broad-leaved weeds are generally less damaging to cereals than grass species with the exception of cleavers which on a single plant basis may be more damaging to yield than wild-oats or black-grass. All species become significant at high densities, and the higher the density the earlier is likely to be the onset of damaging competition. Competition occurs earlier with grass than broad-leaved species, and at high densities grass weeds must be controlled during the autumn or early winter to avoid yield loss. Broad-leaved weeds do not often become seriously competitive before early spring except when crops are sown exceptionally early; however competition can increase markedly during late April and early May, and the likelihood of adverse weather delaying spraying until after crop jointing may well justify the autumn spraying of broad-leaved weeds.

Main annual weeds of spring-sown cereals

The period of annual weed emergence in spring-sown cereals is more restricted than in winter cereals, so that a single correctly timed herbicide application should be sufficient for season-long weed control. Annual weeds normally restricted to spring-sown cereals are knotgrass, redshank, pale persicaria, black-bindweed and common hemp-nettle, while others such as common chickweed, common field-speedwell, the mayweeds, wild radish and annual meadow-grass which are capable of emerging during most months of the year can be problems in both spring- and winter-sown cereals. Cleavers which germinates in the autumn and winter is unlikely to be a problem in spring cereals.

For further information on the periodicity of germination see the BCPC *Weed Control Handbook: Principles*, 1982 7th Edn.

Main perennial weeds of cereals

The main perennial weeds of cereals are the perennial grasses common couch, black bent, and onion couch (false oat-grass) and the broad-leaved species creeping thistle and field bindweed. Their period of main vegetative growth occurs later in the spring than most annual weeds, with flowering commencing during late May or early June. Crop competition is very important in restricting vegetative growth, for perennial weeds will proliferate in thin patches of crop during early summer building up large reserves in their underground root or stem (rhizome) systems. Late vegetative growth will be more significant in early drilled and early maturing crops such as winter barley where the canopy dies back in late June or early July. The reduced crop competition will, with adequate soil moisture, allow perennial weeds to make rapid growth in the few weeks before harvest. The underground root or stem systems occur naturally at varying depths in the soil according to the species. The response to cultivation is linked to the depth of storage organ, so that those species with relatively shallow systems will be discouraged by ploughing. For example onion couch with bulbs virtually on the soil surface has increased with reduced cultivation systems but is readily suppressed by ploughing. Broad-leaved perennials tend to be deeper rooted, and most regrowth arises from below the cultivated layer.

Major Weeds of Cereals

The major weed species of cereals are considered individually in more detail below. The more important grass weeds which have been well researched are dealt with at greater length than the broad-leaved weeds for which there is much less experimental evidence.

Annual grass weeds

Wild-oats *(Avena* spp.*)*. The common wild-oat *(Avena fatua)* is widespread throughout cereal crops in England, and occurs locally in eastern Scotland, and to a lesser extent in Wales, and Northern Ireland. This species can germinate and emerge during autumn winter or spring and although not completely frost hardy occurs in both autumn- and spring-sown cereals.

The winter wild-oat *(Avena sterilis* spp. *ludoviciana)* is more localised within the Midlands and Eastern counties. Germination is restricted to the autumn and winter months so this species only occurs in winter-sown cereals.

Once established wild-oats compete strongly for light, nutrients and water, and if present in large numbers will reduce cereal yields. If wild-oat plants remain green and immature they can slow crop drying and delay and interfere with harvest. Immature plants can contaminate the harvested grain with wild-oat seeds and reduce its value. In addition wild-oat plants can act as hosts of both the cereal cyst nematode and the cereal stem nematode.

Identification and basic biology are outlined in the ADAS Leaflet 452, *Wild-oats*, Ministry of Agriculture Fisheries and Food, Alnwick 1985. Knowledge of the biology of the weed helps in planning strategies for control. The long-term success or otherwise of reducing a wild-oat population will depend on the balance between seed production and seed losses in the life cycle of the weed. Factors which influence the quantity of viable seeds produced and the survival of such seeds to produce plants in future crops are summarised in Table 2.1.

TABLE 2.1 Factors influencing *Avena Fatua* seed production and survival

Seed production favoured by:	*Seed losses* occur through:
Crop competition	Losses from field of origin
* Low crop seed rate	* Time of harvest relative to seed shedding
* Weeds emerging with the crop rather than later	Losses after harvest
Soil type and climate	* Straw burning
* Moist summer weather	* Natural losses influenced by timing of cultivation
* Moisture retentive soils	Losses from the soil
Herbicide use	* Type of cultivation influencing depth of burial
* Poor control	* Age of seed reserves
	Seedling mortality
	* Time of germination relative to seedbed cultivations
	* Time of emergence relative to crop

The numbers of seeds produced by each wild-oat plant will be very variable and will be largely governed by crop competition and by the time of emergence relative to the crop. In experiments averages of 40-50 seeds/plant have been recorded in spring barley crops. In winter wheat 80-100 seeds/plant have been recorded from wild-oats which emerge in the autumn, reducing to 10-30 seeds/plant from wild-oats emerging in the spring when the crop is well established. Seeds which are produced may be lost from the field as contaminants in harvested grain or straw, or they may be destroyed as a result of straw burning or lost naturally from the stubble surface. Seeds which do survive enter the soil to add to the existing reserve of dormant seeds from which subsequent infestations arise. Relatively few seeds in the soil produce seedlings which survive and mature in the next crop. Some will remain dormant but the majority will die directly as a result of rotting or indirectly after unsuccessful germination. The emergence of one seedling for every 10 viable seeds in the soil is a broad generalisation, but this ratio will be influenced by factors such as the age of the seeds in the soil and by the type of cultivation. Old seeds, more than one year old, are less dormant and germinate more readily than seeds of recent origin.

Although wild-oats are widespread in distribution heavy infestations are less common than 10 or 15 years ago due to the repeated use of specific wild-oat herbicides. With high infestations the course of action is straightforward; it is difficult to decide on a control policy for low infestations. Low populations remain a constant threat to modern systems of cereal production because of their potential for rapid build-up and yet the cost of control cannot be offset by higher yields. Many farmers are containing wild-oats at a relatively high cost without eliminating them because of inconsistent herbicide performance. Hand roguing offers a chance to back-up the herbicide with the prospect of eventually dispensing with wild-oat herbicides. With cereal seed crops wild-oats are economically damaging at very low levels, and hand roguing becomes an essential part of control policy.

Precise definition of economic wild-oat thresholds is not possible because of variability in competitive effects which cannot be predicted although a broad generalisation is possible. Management systems have to be sufficiently flexible to accommodate the variations found. An economic threshold of between 5 and 50 wild-oats/m^2 may produce a yield loss equivalent to the cost of application of herbicide. If 10/m^2 is taken as an arbitrary average threshold then if this threshold were not quite reached large numbers could be left causing some immediate yield loss and considerable seed return to carry the problem into the future. Spray decisions based on a population much lower than the economic damage threshold allow for variability of wild-oat behaviour, poor control by the herbicide and also give a margin of safety against a long-term population build-up by restricting seed return. It is necessary to keep populations below 0.5 plants/m^2 (5000/ha) to provide an adequate margin of safety for systems of containment with herbicides, and below 400-500 plants/ha where hand roguing is to be carried out.

A system of population monitoring is needed for such a strategy to be effectively carried out. It is difficult to recognise such low populations of wild-oat seedlings in a cereal crop in order to make immediate decisions on whether or not to spray. Counting panicles in July is easier and more accurate and spray decisons may be made in advance for the next crop. Cereal tramlines provide access for counting panicles at intervals; such a system of monitoring used at the Weed Research Organisation takes about half a man hour/ha to count wild-oat panicles at 7 survey points/ha. A less intensive assessment may be desirable for farmer monitoring but a minimum of 30 survey points should be recorded in each field.

Black-grass *(Alopecurus myosuroides)*. Black-grass is less widespread than wild-oats being mainly confined to the cereal growing areas of central, southern and eastern England, with a recent extension into Yorkshire and the south west of England. It is most common in winter cereals on heavy soils, but is also found on light soils especially where drainage is impeded.

Black-grass has been encouraged by earlier autumn planting, by sowing more winter cereal crops and by the widespread use of reduced cultivations. Black-grass seeds germinate in cool conditions mainly in the autumn, so with early autumn drilling in September or early October fewer seedlings can be destroyed before sowing, and more will successfully establish in the crop. The systems of reduced cultivation associated with early planted winter cereals retain most seed close to the soil surface in the most favourable position for successful germination and seedling establishment.

Black-grass can seriously reduce yields of winter cereals and is probably a greater threat to yield than are wild-oats. The economic threshold for control lies within the range of 20-50 plants/m^2 or 50-100 heads/m^2 depending on the cost of treatment and the efficacy of the herbicide. Losses of about 1 t/ha for every 100 plants/m^2 (producing about 300 black-grass heads/m^2) have been recorded, and populations exceeding 500 plants/m^2 leading to substantial yield losses sometimes occur. As well as reducing yields heavy infestations can hinder harvesting and there is evidence that ergot infestation in black-grass is often associated with ergot in wheat, the same strain affecting both weed and crop.

Heavy infestations produce vast numbers of seeds. Each head contains an average of 100 seeds, and over 100,000 seeds/m^2 have been recorded in cereal crops. Although the viability of this seed is likely to be variable and rarely exceeds 70%, large numbers of viable seeds will reach the ground. Shedding occurs from late June to late August so that in wheat crops most seed is shed before harvest. Winter barley harvested in July may be severely contaminated by seeds which have not been shed before harvest.

After shedding, substantial mortality of seeds occurs, due to fungi, predators and premature germination. This natural mortality of seeds is increased by straw burning. Live seeds on the soil surface become incorporated to different depths according to the cultivation system. Ploughing may raise some old viable seeds to the surface. Most seeds which enter the soil die within three years, but a small but significant proportion survive for many years.

Most seeds have a short but variable period of innate dormancy after shedding and most germination occurs between September and November. Seedlings arising before drilling can be destroyed, but enforced dormancy due to drought or waterlogged conditions may result in a later germination and give more seedlings in the crop. Black-grass plants present before drilling and not destroyed by cultivations or spraying before the crop emerges, are potentially very competitive.

The prolific seed production of black-grass which is especially favoured under systems of reduced cultivations and direct drilling, means that such systems are heavily reliant upon efficient herbicides to limit seed return and prevent populations increasing. A higher level of chemical weed control is needed with reduced cultivations than with ploughing to prevent the weed increasing. Population models based on experimental results show that over 90% control is needed annually to contain a population of black-grass in a system of minimally cultivated winter cereals. Around 65% control with herbicides may suffice with a ploughing system which is less favourable to rapid increase of population.

Long term containment of black-grass in intensive winter cereal cropping systems is dependent on reliable control by herbicides. The reliability of those which are soil acting (see Table 4.4) is reduced by the gradual build up of burnt straw residues in the surface soil associated with reduced cultivation systems. These residues are highly adsorptive to herbicides which can result in inadequate weed control. The problem is very serious in the small number of fields where it occurs because most of the soil acting herbicides are similarly affected and control of black-grass is so heavily dependent on them.

As with wild-oats it is necessary to contain the population at a much lower level than the economic threshold to prevent losses in the long term. This allows for possible failure of the herbicide, field-to-field variation in black-grass competitiveness and changes in the relative costs of herbicides and grain prices. A 'spray decision rule' of 5 plants/m^2 is suggested. Such low numbers of seedlings are difficult to detect. It is more practicable to anticipate this threshold in a succeeding autumn sown crop by making a 'predictive spray decision' based on the presence of two or more heads/m^2 present during June in the previous crop.

Herbicide use alone is unlikely to contain black-grass in winter cereals. Herbicides must be allied with husbandry measures such as straw burning and stubble hygiene so that seedlings are not transplanted into the next crop, and the growing of vigorous competitive crops. With continuous winter cereal cropping containment of populations at low levels is all that can reasonably be expected; eradication is very unlikely with the potentially rapid build-up and the impossibility of hand roguing low populations.

Further information is available in ADAS Leaflet 522 *Black-grass,* Ministry of Agriculture Fisheries and Food, Alnwick 1985.

Brome grasses *(Bromus* spp.*)*. These are traditional hedgerow and field margin species, but since the mid 1970's they have increased dramatically as field problem weeds. Barren brome *(Bromus sterilis)* is the predominant species in cereals, and is found throughout southern England, the Midlands, East Anglia, Lincolnshire and less frequently as far north as east Scotland over a wide range of soil types. Meadow brome *(Bromus commutatus)* is also a localised field problem in Cambridgeshire and surrounding counties.

Seeds of barren brome are produced on panicles with an average of 10-12 spikelets each containing 4-10 seeds. They are shed from late June to early August, the majority having little dormancy and germinating on contact with moist soil. Germination is delayed by inadequate moisture or by exposure to light if seeds remain on the soil surface. Any delay in germination will increase the likelihood of seedlings becoming established in the subsequent crop, especially if the crop is planted early in the autumn. Dry autumns and the trend to drilling winter cereals in September rather than October have contributed to the recent increase in barren brome. With a prolonged autumn drought, a few seeds may persist to infest a following spring cereal but the vast majority of seeds do not remain viable for longer than one year.

In contrast to wild-oats and black-grass, the short persistence of seeds in the soil aids control strategies, so that any break in the annual seed production cycle should dramatically reduce the population. Current herbicides are not reliable enough to prevent seed return entirely, so that the integration of chemical control with husbandry methods is important.

Brome usually originates from the hedge and field margin, and as the seeds are vulnerable to straw burning, a headland firebreak often has the highest level of brome and may require special treatment. Few seedlings are able to emerge from deeper than 13 cm so efficient headland ploughing before or after burning aids control.

Seeds of meadow brome shed somewhat later than barren brome and hence pose a greater threat as a grain contaminant.

An economic threshold population for brome cannot be defined because little work has been done on the effects of brome on cereal yields. However heavy infestations can completely smother cereal crops, so that brome is likely to be equally or more competitive than similar infestations of black-grass.

Further information on these weeds can be found in ADAS Leaflet 777, *Barren brome*, Ministry of Agriculture Fisheries and Food, Alnwick 1981.

Rough meadow-grass *(Poa trivialis)*. Surveys show rough meadow-grass to be present in about 20 % of winter cereal crops, mostly in wheat, and with a more frequent occurrence towards the north and west of the country. Rough meadow-grass is a major constituent of grass swards, and is highly likely to be found in cereals newly established from ploughed out grassland. It is a perennial plant that persists by forming new plants on very short stolons, but which also behaves like an annual in winter crops, growing from seed each year; hence its consideration with other annual weeds. Over-wintering transplants may survive cultivations and produce seeds in spring as well as winter crops. Flowering takes place between June and August, roughly 1000 seeds/plant being produced, most of which are viable with little initial dormancy. Dormancy is induced by burial so autumn cultivations sould be delayed as long as possibly to encourage seed mortality. Seedlings emerge mostly together during late summer or early autumn and plants are shallow rooting, arising from seeds in the top 2 cm of soil. Because of the cold period (vernalisation) requirement for seed production successive spring barley crops can act as a cleaning break provided they follow efficient ploughing to avoid over-wintering transplants. With early-sown winter cereals herbicides have to be relied upon for control.

Rough meadow-grass can act as an alternative host to the fungus which causes leaf spot *(Septoria tritici)* of wheat.

Awned canary-grass *(Phalaris paradoxa)*. This is a relatively new weed first observed in the late 1970s in the coastal regions of Essex and Suffolk. It may have arrived as a contaminant of other species of canary-grass imported as bird seed and often grown as a headland cover crop. It is favoured by winter cropping and by reduced cultivations, has spread fairly rapidly and has been found in some areas of the South and Midlands. The seedling resembles black-grass but can be distinguished by the red sap which exudes from the lower pink stem when crushed. The mature seedhead resembles a timothy spike. One spikelet only in each cluster carries a seed, and the clusters are shed entire, unlike timothy. Shedding starts at the top of the spike generally during mid-July. There is some evidence of seed dormancy enabling awned canary-grass seed to persist in the soil for more than one year. As with sterile brome, infestations are usually confined to the headlands but in some cases have spread out into the whole field. Although the effect on crops is not known this weed grows and tillers vigorously and may be expected to reduce yield significantly.

It is not well controlled by the urea herbicides, which has assisted its initial rapid build-up. Other herbicides can be used with more success and provided early action is taken to recognise and deal with the weed it should not become a serious national problem.

Further information can be found in ADAS Leaflet 915 *Awned canary-grass,* Ministry of Agriculture Fisheries and Food, Alnwick 1984.

Annual broad-leaved weeds

Common chickweed *(Stellaria media).* Chickweed is the most frequently occurring species of broad-leaved weed in arable crops. It germinates throughout the year with distinct germination peaks in the spring and autumn, and thrives in cool wet conditions and on fertile soils. It tolerates low temperatures and readily overwinters producing large competitive plants in the early spring. It is however sensitive to drought. Seeds are produced in large numbers, on average about 3,000 on each plant, although large plants may contain up to 15,000 seeds. Seed germination is variable. Some seeds can germinate immediately they are shed, while others require a period of after ripening before full germination and some remain dormant accumulating viable seeds in the soil. This variability in germination, and the short period between germination and flowering so that two or three generations may be completed in one year, are reasons for the success of this species as a weed.

Cleavers *(Galium aparine).* This is an annual weed of increasing importance in winter cereals. It has a climbing and scrambling habit which means that even low populations form a dense canopy over the crop causing serious lodging and loss of yield. Cleavers is economically the most damaging of the broad-leaved weeds of cereals and individual plants are more competitive than wild-oats or black-grass. 20-30 plants/m^2 have been known to halve wheat yield. Cleavers seed is very difficult to separate from wheat grain, and can be easily spread by sowing contaminated seed. Cleavers has a fairly low temperature requirement for germination (11-15 °C) and germinates mainly in the autumn and early winter although some can emerge in early spring. Seed dormancy and plant vigour are very variable both between fields and within fields, with variation between plants growing in the hedgerow and in the field. High infestations can produce over 20,000 viable seeds/m^2, and individual plants at low infestations over 1,000 seeds/plant. Dormant seeds can persist in the soil for at least 18 months, so with the large numbers of seeds produced there is the potential for rapid increase of population.

Shedding occurs later than black-grass and the majority of seeds are usually still attached to the plants at harvest. Cleavers seed sown with contaminated winter cereals is ideally placed for successful emergence. Seeds will emerge most successfully from the top 5 cm, although some emergence can occur from seeds buried up to 10 cm. Seeds which have shed naturally and remain undisturbed on the soil surface as with direct drilling will germinate less readily than seeds buried by shallow autumn cultivations.

Mayweeds (*Matricaria* spp. and *Chamomilla* spp.). These are annual weeds which can germinate in most months with peaks in the spring and autumn. Scented mayweed or wild chamomile *(Chamomilla recutita)* has finely divided dark green leaves which are pleasantly aromatic when crushed, in contrast to Scentless mayweed *(Matricaria perforata)* which is not aromatic. Both species produce abundant seed which can survive and remain dormant in the soil for many years. Mayweeds tend to remain green and competitive throughout the life of the cereal crop in contrast to many other broad-leaved weeds which die away before cereals ripen. Pineappleweed *(Chamomilla suaveolens)* is not invasive in well grown crops but persists in gateways and compacted areas. It is distinguished from the other mayweeds by having a prominant domed yellow receptacle with no white petals.

Speedwells (*Veronica* spp.). Ivy leaved speedwell *(Veronica hederifolia)* germinates in the autumn and winter and is confined to winter cereals. The large cotyledons resemble cleavers, but have a small point at the tip instead of an indentation, and the true leaves are ivy shaped. Common field speedwell *(Veronica persica)* which germinates throughout the year mainly in the spring and autumn is also a weed of spring cereals. Both species are not controlled by urea type herbicides and in recent years have increased in importance particularly in winter cereals. They are readily controlled by HBN/mecoprop mixtures.

The Polygonums: Knotgrass *(P. aviculare);* **Black-bindweed** *(P. convolvulus)* (alternatively *Fallopia convolvulus);* **Redshank** *(P. persicaria);* **Pale persicaria** *(P. lapathifolium).*

Knotgrass has a low temperature requirement for germination and emergence usually begins in February, reaching a peak in March or April so that this species is mostly confined to spring sown cereals. The plant can produce dense mats of tough stems, hence the alternative name wireweed. Flowering occurs between July and October and the seeds which are dark brown and triangular can be very persistent in the soil.

Black-bindweed germinates about a month later than knotgrass starting in March with the main flush in April and May. It is a climbing weed and may cause crop lodging. Each plant produces about 200 shiny black seeds which need a period of chilling before germinating.

Redshank and pale persicaria are also weeds of spring cereals germinating over the same period as black-bindweed. They are most abundant on wet soils.

Field pansy *(Viola arvensis)*. This weed is increasingly common in winter cereals, and is often difficult to control with herbicides. Although generally considered to be not very competitive, high populations have built up in some situations posing a threat to crop yield. The seeds can germinate in both autumn and spring and the protracted emergence makes timing of effective control measures difficult. The plant flowers from April to October and many seeds are dispersed by being violently ejected from capsules when ripe.

Perennial weeds

Common couch *(Elymus repens)*; **Bent grasses** *(Agrostis* spp.*)*. Couch grass is a term used to include a number of creeping grass weeds. The most important of these are the rhizomatous species, common couch *(Elymus repens)* and black bent *(Agrostis gigantea)*. Common couch may be distinguished at an early vegetative stage by the presence of auricles clasping the stem at the base of the leaf blade, and later by the distinct flowering spike. This is similar to perennial rye-grass except that the spikelets have their flat face opposite the stem, very different from the open, branched flowering head of the bent grasses. Creeping bent *(A. stolonifera)*, with surface creeping stems, brown bent *(A. canina)* and common bent *(A. tenuis)* with short rhizomes, are less important than black bent as cereal weeds.

Viable seed production by which new plants and rhizome systems may develop is much greater in black bent than in common couch. The seed appears to have little initial dormancy, but dormancy may be induced later in buried seed. Seedlings are not very vigorous and are susceptible to disturbance. The most important means of propagation is from existing rhizomes, an aspect which has been extensively studied. Growth from over-wintered rhizome is comparatively slow in the spring, the emergence and growth of shoots being slower than those of spring sown cereals. New rhizome growth commences after the aerial shoots have started tillering, reaches a maximum in early to mid- summer and continues at a lower rate during the autumn. Crop competition will suppress aerial growth and delay the formation of new rhizome, but common couch will normally spread its flag leaf above the crop and continue growing after crop ripening has commenced. It has been shown that common couch may double the weight of rhizome present in undisturbed cereal stubble between early September and November. Black bent flowering shoots ripen and die back before cereal harvest but the basal leaves persist to feed the growing rhizome system which will continue to develop in undisturbed stubble.

Further information can be found in ADAS Leaflet 89 *Couch*, Ministry of Agriculture Fisheries and Food, Alnwick 1985.

Common couch and the rhizomatous bent grasses are hosts of the fungus which causes take-all disease of wheat and barley.

Onion couch *(Arrenatherum elatius* var. *bulbosum)*. This is a perennial weed form of false oat-grass found commonly on roadside verges and field margins, propagating by strings of bulbs at the stem bases. The bulbs are readily separated by cultivation and each one is capable of growing into a separate plant. It also reproduces by seed, some of which will give rise to plants of the bulbous form. It is an increasing problem in cereals being favoured by minimum tillage and in particular by direct drilling. In these situations control is dependent on herbicides. It seldom becomes a problem where regular ploughing is carried out.

For further information see ADAS Leaflet 89 *Couch*, Ministry of Agriculture Fisheries and Food, Alnwick 1985.

Creeping thistle *(Cirsium arvense)*. One of the most persistent and troublesome perennial weeds of both grassland and arable land is creeping thistle. It is a highly competitive weed of cereal crops, heavy infestations of about 30 shoots/m^2 having been shown to halve cereal yields. It is considered to be three to four times as competitive as wild-oats, and about three times as competitive as an equivalent ground cover of bindweed.

Creeping thistle spreads vegetatively by means of an underground root system consisting of vertical roots which can penetrate to a considerable depth, and lateral roots from which new shoots arise, mostly in the spring and autumn. Creeping thistle can spread rapidly. The roots fragment naturally and are easily broken by cultivations, but control by cultivation is not easy. Each piece invariably produces a new plant with new shoots arising a few days after fragmenting. Shoots generally emerge over a period late in the spring, often after the normal time for spraying weeds in cereals. Spread by seed is less important than vegetative propagation. Seeds readily fall from the hairy pappus (thistledown) and are not carried far. Seeds can persist in the soil for many years maintaining a thistle patch.

Field bindweed *(Convolvulus arvensis)*. Field bindweed is a widespread perennial weed of arable land which can be very persistent and difficult to eradicate. Although less competitive than creeping thistle, there is evidence that a 50% ground cover at harvest will have reduced the yield of wheat by 10% in a standing crop. The climbing habit of bindweed leads to lodging and so increases harvesting problems as well as reducing yield.

Bindweed has a perennial root system consisting of deeply penetrating vertical roots and shallower horizontal laterals capable of sending up new shoots. The main period of shoot growth is from June to September so cereal herbicides sprayed at the normal time cannot be relied upon for control. Cultivation after harvest produces root fragments and although these can produce vigorous shoots, root production is poor and they usually perish. Successful regrowth after cultivation comes from buds on the roots beneath the depth of cultivation. Food reserves are concentrated in the root system below the cultivated layer so that depletion of reserves by cultivation takes place only very slowly. Deep roots can persist for three years or more without the reserves being replaced; stored food reserves give the roots a high potential for regeneration.

Disease Problems in Cereals

General considerations

The list of pathogens attacking UK cereal crops includes three viruses and a few relatively unimportant bacteria, but by far the majority of the diseases to which cereals are subject are caused by parasitic fungi. Roots, stems, leaves, glumes, awns and developing seeds may all be attacked, and the plant is vulnerable from the time the seed germinates until the green tissues finally die during the grain ripening period.

In the absence of cereal crops the disease causing organisms survive on root debris in the soil (e.g. take-all), on the leaf and stem debris on the soil surface (e.g. eyespot, rhynchosporium), on volunteer plants (e.g. mildew and rusts), on weed and hedgerow grasses (e.g. barley yellow dwarf virus (BYDV), take-all and ergot), or even, occasionally, on alternate hosts of completely different plant families (e.g. some rusts). In addition, a number of diseases are seed-borne, the fungi which causes them being carried either on or near the seed surface (e.g. bunt, barley leaf stripe) or deep within the embryo (e.g. loose smut). Spread within and between fields may be by wind blown or rain splashed spores, by insects (BYDV and the "honeydew" stage of ergot) or by infected soil moved by cultivation practices or on contaminated machinery, e.g. barley yellow mosaic virus (BaYMV).

The fact that a pathogen is present in a field does not mean that the crop will necessarily suffer seriously from its depredations. Disease severity will depend on such factors as weather conditions, micro-climate within the crop, varietal susceptibility and, to a lesser extent, the nutrient status of the plant and soil conditions. The following notes briefly describe the more important diseases of the cereal crop and give some indication of the conditions which favour their development.

Important diseases

Take-all *(Gaeumannomyces graminis* var. *tritici).* Take-all is a soil-borne fungus occurring naturally on the roots of many wild grasses and present in most soils where cereals are grown in temperate climates. Wheat and barley are both very susceptible, but the disease causes less serious losses in barley than in wheat and spring sown crops suffer less severely than do winter cereals. Rye shows a measure of resistance to the disease and triticale is intermediate in susceptibility between wheat and rye. Oats are immune to the commonly occurring *G.graminis* var *tritici*. There is another form of the fungus *(G. graminis* var *avenae)* which will attack oats as well as the other cereals, but fortunately this form is rare in Britain occurring only in some western areas. Infection by the take-all fungus causes a black rot of the roots and stem bases. Affected plants may be very stunted and poor if infection occurs early, later infection has less effect on the vegetative growth of the plant, but causes premature ripening with "white heads" and shrivelled grain. Disease levels in the first cereal crop after non-cereal break are generally low (unless fungus has been carried through the break on perennial grass weeds). Disease builds up in cereal sequences reaching a peak after two, three or four cereal crops, thereafter its severity declines and yield losses are generally less than they were in the peak take-all years. The cause of the so called "take-all decline" (TAD) is not fully understood, but it appears to be due to the build-up of microbial antagonists in the soil. The disease tends to be more severe on light alkaline soils where the hyphae of the fungus are able to run freely over the roots of the young plants. Severe attacks can also, however, occur on heavy land especially if bad drainage or a soil structural problem limits rooting thus accentuating the effects of root loss caused by the disease. "Acid patch" take-all occurs where the pH level is low enough to restrict rooting, but not so low that the fungus is inhibited (as happens below pH 4.5). Plants deficient of potash or, more particularly, of phosphate are particularly vulnerable to attack by the pathogen. So long as neither of these major nutrients is limiting, extra nitrogen in the spring helps the plant to cope with the disease by enabling fresh root production to replace roots lost through the pathogen. Early-sown crops are very susceptible to the disease and in such crops take-all patches may show up in late autumn. A mild winter followed by a spring in which soil moisture levels are high favours the development of the disease; the final expression of symptoms in premature ripening, however, is most severe when summer drought places diseased plants under severe moisture stress.

Eyespot *(Pseudocercosporella herpotrichoides)*. Two forms of *P. herpotrichoides* exist, one attacking wheat and barley only, the other, wheat, barley and rye. Oats may also be attacked but seldom suffer severely. The fungus survives on stem bases in the soil, and a non-cereal break of at least two years is necessary to reduce inoculum levels effectively. During the autumn, winter and early spring spores produced on straw fragments at the soil surface are dispersed by rain splash and infect the stem bases of the young cereal plants. The symptoms are usually first seen in early spring as diffuse brown smudgy areas on the lower leaf sheaths. Later the lesions assume the typical lens shape with a dark border and often with a "pupil" of black dots in the centre; a leaf sheath often splits lengthwise down the lesion. The symptoms are more readily seen on wheat than on barley where high pigmentation of the leaf sheaths in early spring, the frequent presence of rhynchosporium lesions and browning caused by *Fusarium* spp. can make diagnosis very difficult. As the disease progresses the fungus penetrates through the leaf sheaths to the base of the stem where, under favourable conditions, it may girdle the stem base causing premature ripening and poorly filled ears. Straws infected in this way are often so weakened that they may break to cause straggling or lodging. The disease tends to be particularly severe in thick crops on the heavier, wetter soils. Early-sown crops are exposed longer to infection and are liable to be particularly severely attacked. Spring-sown crops tend to escape severe infection.

Sharp eyespot *(Rhizoctonia cerealis)*. This soil-borne fungus attacks the stem bases of wheat, barley, oats and especially rye, causing lesions very similar to those caused by the eyespot fungus *(P. herpotrichoides)* . Sharp eyespot lesions are, however, generally more sharply defined at their margins than are those of eyespot and are somewhat asymmetric. Instead of the black "pupil", at the centre of an eyespot lesion purple brown fungal masses (which can readily be scraped off) develop on sharp eyespots. The centre of the lesion is usually pale buff-brown in colour and tissue shredding may occur. The stems infected with the sharp eyespot fungus often bear multiple lesions, sometimes to a height of 30 cm or more above soil level. Where infection is severe small resting bodies, at first buff and later black, develop on the lesions or within the stems.

Sharp eyespot tends to be particularly severe on light soils and in early sown winter cereals. Infection is favoured by slow plant growth during the winter and spring. The disease does not necessarily increase with repeated cereal cropping and severe attacks can occur after leys or non-graminaceous crops.

Barley stunt disorder *(Rhizoctonia solani)*. Related to, but distinct from, the fungus that causes sharp eyespot *(R. cerealis)*, *R. solani* is a pathogen of light land cereal crops, especially of barley. It attacks the young roots in such a way that they become constricted at intervals along their lengths to appear, in severe cases, rather like strings of sausages. Ultimately the root becomes completely severed at one of the constrictions and as laterals are then produced above the point of severance the root system takes on a very "fangy" appearance. Symptoms may become obvious in autumn, winter or early spring, affected plants being thin and stunted and their leaves often showing a purplish tinge. Patches of such stunted plants are often kite shaped, their long axes being in the direction of the primary cultivation. Better growth may occur in the centre of the patch.

Although it can occur where cereals are grown continuously, barley stunt is often most severe after old grassland. The disease is favoured by direct drilling and minimum cultivation techniques, and may be reduced by ploughing.

Fusarium species: seedling blight, brown foot rot and ear blight. The *Fusarium* species causing this complex of diseases may be seed-borne or soil-borne. Infected seedlings may be killed before or soon after emerging. Those that survive may exhibit browning of the stem bases and root systems. Patches of diseased plants, stunted and with pale leaves, may occasionally be encountered in wet autumns, but the symptoms more frequently develop during the spring and summer. Severely infected plants show a brown rot of roots and stem bases, this can lead to premature death at any time between ear emergence and harvest. The "white heads" symptom is sometimes confused with that caused by take-all or eyespot, and to make diagnosis more difficult two or three of these diseases may occur on the same plant. Typically fusarium infection causes the lower internodes to appear brown and water-soaked with the discoloration being particularly intense just below the node. On the node itself sporulation of the fungus may produce a pink encrustation of spores. This foot rot symptom is usually most severe during or after periods of spring and early summer drought. Later in the season infection may progress up the stem by rain splash of spores and diffuse necrotic areas develop on the upper leaves. Where rain occurs after anthesis the fungi may spread to the ears killing whole spikelets or complete sections of the ear. The dead glumes become bleached and in wet weather covered by mycelium and/or pink spore masses of the fungus. The grain within the dead glumes is shrivelled and poor.

Seedling blight most frequently occurs under dry soil conditions. Brown foot rot is favoured by high humidity around the stem bases. Severe attacks are most likely to occur on heavy soils; the symptom expression (white heads) may be favoured by drought.

Snow mould *(Fusarium nivale).* While the Fusarium species most frequently associated with brown foot rot *(F. culmorum* and *F. avenaceum)* are favoured by warm, dry soil conditons, attacks by *F. nivale* tend to be more severe after cold winters. The fungus is particularly aggressive under the conditions provided by prolonged snow cover. In such circumstances it will kill leaves and sometimes whole plants, the dead tissues often becoming covered with a white mould growth. Less aggressive attacks during the winter period are often characterised in the early stages by browning of the leaf sheaths and the lower few millimetres of the leaf blade (*Rhizoctonia cerealis* can sometimes produce a very similar symptom). Snow mould can occasionally cause serious losses in those areas of the country where winters are most severe.

Snow rot *(Typhula incarnata)*. A common parasite of winter cereals, particularly of winter barley, grown on light land is *Typhula incarnata*. A soil-borne fungus, it will attack both cereals and grasses and, in the absence of host plants, it can survive in the soil as resting bodies known as "sclerotia". The sclerotia, red-brown in colour and about the size of a clover seed, can be found from late winter onwards on the underground parts (roots and lower leaf sheaths) of affected plants. Surviving the death of the host they germinate in autumn or winter to produce club shaped "basidiocarps", spores which infect the current season's cereal crop.

Sclerotia may often be found on symptomless plants – the pathogen appearing to cause its host little damage. Under certain circumstances however, more particularly under prolonged snow cover, a brown rot of the stem bases occurs. The leaves turn yellow and wilt and death of individual tillers or of complete plants can ensue.

Compensatory growth by surviving tillers or surrounding healthy plants can minimise yield losses if only occasional tillers or plants are killed. If, however, large areas are affected, or if a second growth check (e.g. spring drought) is encountered before such compensation has occurred, more substantial losses may be sustained.

Septoria leaf spot and glume blotch *(Septoria* spp*)*. Wheat, barley, oats and rye are susceptible to a range of a *Septoria* species which can attack leaves, glumes or, occasionally, stems. The most damaging of this group of pathogens, however, are *S. nodorum* and *S. tritici* on wheat. Both of these will survive on diseased stubble, *S. nodorum* is also carried on seed and can occasionally cause seedling blight.

The leaf spot caused by *S. tritici* usually becomes conspicuous in late winter when pale brown to greenish grey oval lesions dotted with the small black spore cases of the fungus, develop on the lower leaves of wheat. Early sown crops in western, or eastern coastal, areas are particularly prone to attack. As the disease spreads the lesions increase in size often joining up to affect large areas of leaf. Although on very susceptible varieties *S. tritici* can remain active throughout the season, other varieties often grow away from the disease as temperatures rise in early summer. If, however, the summer is wet leaf symptoms of *S. tritici* are soon replaced by those of *S. nodorum* which are more lens-shaped and slightly browner in colour with a yellow margin. The spore cases of *S. nodorum* are far less prominent than those of *S. tritici* being pinkish brown, embedded within the tissue, and often detectable only when the leaf is held against the light and viewed through a hand lens. *S.*

nodorum infection may spread to the ears where purplish brown lesions develop on the glumes.

Leaf blotch *(Rhynchosporium secalis)*. This will attack rye, but is most commonly encountered as the cause of the common leaf blotch of barley. The pathogen survives on stubble debris from previous crops or on volunteer plants. Seed-borne infection may play a part in introducing the disease into new areas. Spores produced on leaves or dead leaf debris are spread in water droplets especially during rain. The disease is favoured by cool (10-15 °C) and wet conditions. In the west of England symptoms may develop in the late autumn, but in the east the disease seldom becomes obvious before late winter to early spring. Pale green, oval, water-soaked lesions develop on the leaf blade or sheath. These enlarge slightly with age and become pale brown with a dark brown or purplish brown margin. Very frequently lesions develop at the base of the leaf and this can be especially damaging as a lesion in this position will quickly kill the whole leaf. Later in the season, small brown-bordered lesions may develop on the glumes.

Net blotch of barley *(Pyrenophora teres)*. While seed-borne infections may be important as the means by which the disease is introduced into a new area, the main source of *P. teres* is infected straw debris and stubble from previous infected crops. Spores produced on this infected debris and dispersed by wind or rain splash serve to spread the disease to plants emerging in the near vicinity. While spores are less readily dispersed by air currents than are those of rust or mildew, some field-to-field spread can occur.

On young plants infection typically produces lesions made up of a network of narrow brown lines, but less distinctive lesions (lines or spots, often surrounded by yellow chlorotic areas) may also be produced. Later in the season a stripe symptom tends to dominate. Stripes may be up to 5 cm long, typically they are light brown with darker brown lines running at right angles to the length of the stripe. In wet weather disease development may be very rapid causing premature death of the leaves. Infection of awns and glumes may also occur.

Although it had always been fairly easy to find in barley crops, net blotch seldom caused serious yield losses until the late 1970s. During 1979 and 1980, however, severe infections occurred in most areas of England, infections being particularly severe on winter barley in the south and west. This sudden upsurge of net blotch occurred in seasons when wet weather provided conditions conducive to the development of the disease. These seasons happened to come at a time when the acreage of winter barley, and particularly of the very susceptible variety Sonja, was rapidly expanding. A move towards ever earlier sowing dates, increasing use of minimal tillage techniques, which left infected debris on the soil surface, and an increase in nitrogen usage which rendered plants more susceptible to infection also contributed to the upsurge of the disease.

Leaf stripe of barley *(Pyrenophora graminea)*. Unlike *Pyrenophora teres*, *P. graminea* is entirely seed-borne. First signs of infection take the form of narrow stripes running the entire length of the first leaf. The stripes are at first pale green, later becoming yellow and finally brown. As subsequent leaves emerge, they too become infected so that most leaves on any infected plant will show symptoms. Often the tissues split along the stripes giving the leaves a shredded appearance. If growing conditions are particularly favourable the ear may escape infection, but more frequently it is invaded by the fungus. Affected ears are usually deformed, frequently failing to emerge completely from the sheath and the grains do not fill. Spores produced on the affected leaves will not infect other leaves, but will invade the ears of neighbouring healthy plants, thus increasing the level of inoculum in the seed produced by the crop.

Cephalosporium leaf stripe *(Cephalosporium gramineum)*. While it can affect other cereals and grasses cephalosporium leaf stripe is primarily a disease of wheat. *C. gramineum* is a soil-borne fungus which gains access to the plant through damaged roots and then grows up into the vascular tissues of the shoots. The fungus blocks the water-conducting vessels restricting the passage of water and nutrients to the leaves and leading to the development of broad yellow stripes running from the base of the sheath to the tip of the blade. If the node is cut across the vascular strands leading into the yellow areas often show brown discoloration. Yields of affected plants may be substantially reduced, but the disease is seldom severe except after old grass where wireworm damage to the roots facilitates its entry. Cephalosporium leaf stripe is one of the few diseases which is definitely aggravated by direct drilling and by straw incorporation.

Powdery mildew *(Erysiphe graminis)*. Wheat, barley, oats and rye are all affected by powdery mildew, but specialised forms of the fungus exist, each being restricted to its own species (*E. graminis* f. sp. *tritici*, being restricted to wheat, *E. graminis* f. sp. *hordei* to barley and so on). Within each special form there are very many physiologic races, each adapted to particular cultivars of the host species.

Mildew first appears on the plants as white fluffy pustules on the leaf blades or sheaths (in wheat the glumes may also be affected). The fungal growth is superficial and can readily be rubbed off. When conditions are unfavourable, or the variety resistant to attack, the fungus may cause small dark brown flecks on leaves with little fungal growth on the surface. The fluffy white pustules produce masses of vegetative spores which may be transported for considerable distances in wind currents. With age the pustules become matted and brown and vegetative spore production declines. From mid-summer onwards, black spore cases (cleistothecia) may be found embedded in the felted pustules. These cleistothecia are produced as the result of a sexual fusion, they survive the death of the plants and, with the first rains of autumn, release "ascospores" which infect volunteer plants or early emerging autumn sown crops. The disease overwinters on green plant tissue, spreading from autumn-sown to spring-sown crops during the following season.

The development of mildew is favoured by warm (10-20 °C) sunny weather with sufficient wind to disseminate the spores. Temperatures above 25 °C check spread. Rapidly growing crops are most susceptible, and susceptibility increases with increasing levels of available nitrogen.

Black stem rust *(Puccinia graminis)*. In the late 18th and early 19th centuries, black stem rust was obviously a very important pathogen, for records show that at that time farmers were assiduously destroying the barberry bushes which acted as its alternate host. Nowadays black stem rust is seldom encountered except in the far south west and Wales and even there it rarely causes significant yield losses.

The disease is first noticed as red-brown pustules on the leaves, leaf sheaths and stems of cereals. These pustules contain "uredospores" which are carried by the wind and serve to spread the disease during the summer months. Later in the season large black pustules which give the disease its name are produced. These contain teleutospores, which survive the winter on straw debris and germinate in the spring to produce a third kind of spore (basidiospores) which will infect only barberry. The sexual stage of the fungal life history takes place within the barberry leaf, and as a result of this bright orange "aecidospores" are formed, these carry the infection back to cereal crops.

On a world-wide scale black stem rust is one of the most important cereal diseases, in this country however, it is so uncommon that specific control measures are not necessary.

Brown rusts *(Puccinia reconditia* and *P. hordei)*. Special forms of *Puccinia reconditia* affect wheat and rye; *P. hordei* affects barley only. The disease first appears as orange-brown pustules on the leaves and leaf sheaths. On barley the pustules are often surrounded by a distinct yellow halo. The wind-borne uredospores produced in the pustules serve to spread the disease during the summer. Later the pustules turn black as the teleutospores develop. These survive the winter on straw debris, germinating in the spring to produce basidiospores, which will infect, not cereals, but only the alternate hosts of the fungi (star of Bethlehem in the case of *P. hordei*, and in the case of *P. reconditia* various weeds including meadow rue and bugloss). In this country the alternate hosts are of little importance as both pathogens can survive the winter in the uredospore stage on volunteer plants and, more particularly, autumn-sown cereals.

Although brown rust can be found in late autumn and winter in autumn-sown crops, severe attacks do not usually occur until later in the season – mainly after ear emergence. At this time brown rust can build up very rapidly if weather conditions are favourable – i.e. day temperatures of 15-22 °C with dew formation at night.

Yellow rust *(Puccinia striiformis).* Specialised forms of *Puccinia striiformis*, each restricted to its own host, affect wheat and barley. Affected leaves show small orange-yellow spore pustules which characteristically occur in long lines giving the disease its alternative name of "stripe rust". The "uredospores" produced in these pustules are wind-borne and serve to spread the disease both within and between the crops. From mid-summer onwards, black "teleutospore" pustules appear, but these seem to be functionless as no alternate host of *P. striiformis* is known. The fungus survives harvest on late green tillers and volunteer plants and overwinters on volunteers and autumn-sown crops. Autumn infection usually produces discrete foci of infected plants which may not become apparent until the spring. Cool (10-15 °C) wet weather in early summer favours disease development and during such conditions epidemic spread occurs from the first formed foci. A spell of warm (20-22 °C) dry weather will check an epidemic.

Early sown crops in which infection becomes established in the autumn suffer particularly from the disease. Risks of autumn infection can be reduced if infected volunteer plants in nearby stubbles are ploughed under or sprayed-off before autumn-sown plants emerge. Excessive use of nitrogen fertilisers renders the plants particularly susceptible to attack.

Crown rust of oats *(Puccinia coronata).* Special forms of *Puccinia coronata*, each specific to its own host, will infect oats and various grass species (including cultivated grasses). No other cereal species is infected. Orange pustules of wind dispersed uredospores appear on the leaf blades and sheaths; occasionally the stem and panicle may also be affected. Later in the season black pustules of teleutospores appear. Teleutospores survive on straw debris until the spring when they germinate to produce basidiospores which infect the alternate hosts of the pathogen – buckthorn and alder buckthorn. Aecidiospores produced on the alternate hosts carry infection back to the oat crop. In fact in most parts of the country the rust can overwinter in the uredospore stage on volunteer plants and autumn-sown crops, so the alternate hosts do not play a particularly important part in the epidemiology of the disease.

Crown rust is favoured by warm and wet or humid weather. Severe attacks do not usually occur until late in the season, but severe infection during the period from panicle emergence to the milky ripe stage can seriously affect yields, the grain produced being light and shrivelled.

Late season diseases of leaf and ear. During anthesis the upper leaves and glumes of cereal plants become liberally dusted with pollen – a nutrient rich food base which can readily be colonised by weakly parasitic fungi such as *Botrytis cinerea* and *Alternaria alternata*. If conditions are warm and humid these fungi advance from their "base camps" on the pollen grains into the tissues of the leaves themselves. Diffuse necrotic lesions develop especially on the "bend" of the leaf and, by reducing leaf longevity, can adversely affect grain size. On wheat ears lesions caused by minor pathogens such as *B. cinerea* and *Fusarium poae* are often confused with those caused by more virulent parasites such as *Septoria nodorum* and *Fusarium culmorum*. The following notes may aid diagnosis:-

• *Botrytis* lesions begin as a pale tan discolouration of the upper part of the inner glumes, extending back from the point where the old anthers are trapped within the spikelet. Later the lesion is normally found to have a rather diffuse purplish margin and a pale (almost white) centre. Infection often extends to involve the whole of the lemma and palea and the developing seed may be attacked and killed. A grey felt of sporulating fungus is often found on the dead glumes. (Botrytis is also a frequent cause of spikelet death in barley.)

• *Fusarium poae* causes a smaller, more discrete lesion than *Botrytis*, with a straw-coloured centre bounded by a very distinct purplish brown border. Again the lesions are almost always associated with old anthers and are found on the inner glumes only. The developing seeds are not normally affected. (*Rhynchosporium* can cause somewhat similar symptoms on the glumes of barley.)

• *Fusarium culmorum* causes genuine **ear blight**, whole spikelets or larger sections of the ear being affected. In the early stages the affected spikelets have a dull, purplish appearance. Where they join the main stem the tissue will be brown and a crust of orange spores may often be found at the very base of the spikelet stalk. Later the spikelets become completely bleached and may be covered with white mycelium and pink or orange spores.

• *Septoria nodorum* causes purple-brown blotches of both inner and outer glumes; pycnidia may be present but are very difficult to see even with a hand lens. In very humid conditions their position may be pin-pointed by the presence of tiny pink spore tendrils on the glume surface.

• *Brown smudging* of the outer sterile glumes is sometimes the result of minor pathogens invading tissue damaged by thrips.

In wet seasons the glumes (especially those of ears which have died prematurely as a result of, say, an attack of take-all) may become covered with the sooty moulds *Cladosporium herbarum* and *Alternaria alternata*. While the late build-up of these fungi has comparatively little effect on yield, they may reduce quality. *Alternaria*, in particular, can cause the condition known as "black point" a dark discolouration of the embryo end of the seed.

Bunt of wheat *(Tilletia caries)* and **covered smut of barley** *(Ustilago hordei)*. Bunted wheat ears are difficult to distinguish from healthy ears except in that the glumes are pushed further apart by the rounded bunt balls which replace the grain. Barley ears with covered smut are more readily recognised being grey black in colour and bearing only shrivelled awns. In both cases the seed is replaced by a mass of black, and in the case of bunt, evil-smelling, spores contained within the seed coat. At threshing the seed coat ruptures and the spores are released to contaminate healthy grain. When this contaminated grain is sown the spores germinate, the fungus infects the emerging seedling and grows with the plant until after ear emergence when it invades the developing grains, replacing their substance with masses of its own spores. Because they are carried superficially on the seed both *T. caries* and *U. hordei* are easily controlled by seed treatment.

Loose smut of oats *(Ustilago avenae)*. This loose smut has a life history similar to those of *T. caries* and *U. hordei* and may be controlled in a similar manner.

Loose smut of wheat and barley *(Ustilago nuda)*. Unlike those of the wheat bunt and barley covered smut fungi, the spores of *U. nuda* are not contained within the seed coat, but within a very delicate membrane which ruptures soon after the emergence of the ear allowing the powdery spores to blow away. Carried by the wind to the feathery stigmata of surrounding wheat or barley plants the spores germinate and the fungus grows down into the ovary finally establishing itself in the embryo of the developing seed. Seed infected in this way develops and germinates normally, but the fungus grows up with the plant finally invading the ear and converting the spikelets into further masses of black spores. Because it is carried within the embryo rather than on the seed surface *U. nuda* is more difficult to control than are the covered smut fungi (*Tilletia caries* and *Ustilago hordei*).

Ergot *(Claviceps purpurea)*. Wheat and rye, occasionally barley and, very occasionally, oats are prone to attack by ergot. It is common on many species of wild and cultivated grasses and the strains infecting some of these (notably black-grass and meadow foxtail) will also infect wheat. "Open flowering" species such as triticale and some modern winter barley varieties are particularly vulnerable to attack.

The most obvious symptom of the disease is the formation of purplish-black "sclerotia" (which may be up to 2 cm long) in place of some of the seeds in the ear. The sclerotia may fall to the ground or be harvested with the grain and be sown with seed corn the following year. They germinate in the spring, each producing from one to fifty fructifications up to 2.5 cm long and terminating in globular heads which give them the appearance of miniature drum sticks. Low temperatures are necessary to vernalise the sclerotia; germination is favoured by warm (12 °C) wet weather.

Spores extruded from flask-shaped "perithecia" embedded in the swollen heads of the fructifications are carried by air currents to the flowering heads of cereals or grasses. Caught on the feathery stigma a spore will germinate and grow down to penetrate the ovary within 24 hours. Within 5 days a drop of sweet, amber "honeydew" may be found extruding from the glumes of the affected floret. Contained within the droplet are spores of a second kind, "conidia", which (spread by insects, wind or rain splash) serve further to build-up the disease within the crop. Ergots germinating in May can infect black-grass, and conidial spread in the weed grass can, in some seasons, build up high levels of inoculum in infested fields before wheat flowers in June. In time conidial production gives way to the production of sclerotia and the cycle is complete.

Ergot sclerotia contain toxic alkaloids poisonous to both humans and cattle (though they have found positive use in obstetrics). Stringent regulations control the levels of ergot allowed in grain for seed, feed or milling.

Barley Yellow Dwarf Virus (BYDV). Infection by the barley yellow dwarf virus (BYDV) causes bright yellow discolouration of the leaves of barley, the discolouration beginning at the tip eventually extending to the base of the leaf. In wheat affected leaves often have a pink or purplish tinge and in oats they are bright purple. Early infection can cause marked stunting of plants and substantially reduce their yield. Wild and cultivated grasses may also be infected and can act as reservoirs of infection for nearby, or succeeding, cereal crops.

The virus is carried by various aphid species, more particularly the bird cherry aphid *(Rhopalosiphum padi)* and the grain aphid *(Sitobion avenae)*. The bird cherry aphid is the vector usually responsible for introducing the severe strain of the virus into early sown winter cereals during the autumn. Winged aphids coming into the crops during September or October serve to infect individual plants, and their wingless progeny spread the virus outwards from these primary points of infection until their activity is checked by the onset of cold weather. It is in these more or less circular, autumn established foci of infection that the most serious yield losses occur.

While the bird cherry aphid is also responsible for some spread of the disease in the spring, it is the grain aphid which is the main vector during the summer. Late infections may cause striking discolouration of the flag leaves but they cause comparatively little reduction in yield.

Barley Yellow Mosaic Virus (BaYMV). First found in the UK in 1980 and now widespread in the south and east of England, it affects only autumn-sown barley. Symptoms appear in late winter when patches of pale plants may be found in affected fields. As the surrounding crop grows away plants in the affected areas remain backward and stunted. Closer examination shows them to have a spiky appearance, their leaves being inrolled and mottled, with pale green and dark green tissue producing a streaky mosaic. The mosaic is usually clearly visible even on the youngest unexpanded leaf. Later the leaves become speckled, especially near their tips, with brown necrotic spots. As temperatures rise the mosaic symptoms are lost but affected plants remain stunted and thin with fewer and smaller grains/ear. In the more susceptible varieties yield losses can be substantial.

The virus is carried by the soil borne fungus *Polymyxa graminis*. Cultivation practices serve to spread the disease within infected fields; rigid tine implements which drag root debris considerable distances (as compared with mould board ploughs which simply bury it) provide a particularly effective method of spreading infection. Spread between fields may be by means of contaminated soil on implements and vehicles, or on the feet of men, animals or even birds.

Varieties differ in their susceptibility to the disease, but the varietal picture is made more complicated by the apparent existence of at least two distinct strains of the pathogen.

Identification of cereal diseases

Identification may be assisted by reference to the Disease Identification Card series (coded IC/1 to IC/19) which are available from MAFF Publications, Ministry of Agriculture Fisheries and Food, Alnwick.

Pest Problems in Cereals

General considerations

In England and Wales cereal crops may be attacked by some 36 groups of pests belonging to such diverse animal taxa as insects, myriapods, mites, nematodes, molluscs, birds and mammals. Those most commonly encountered are cereal aphids, leatherjackets, frit fly, wheat bulb fly, wireworms, cereal cyst nematode, stem nematode, field slug, various birds and mammals.

Not all cereal species are equally susceptible to the above range of pests. Thus wheat, barley and rye are attacked by the stem-boring larvae of wheat bulb fly whereas oats are immune to this pest. Again, stem nematode attacks oats and rye but not wheat or barley in this country.

Many cereal pests naturally inhabit grassland, which provides a permanent reservoir from which mobile pests can migrate to attack cereal crops. 'Ley pests' are a group of insects that feed on grasses and which transfer from the ploughed sward to the following cereal crop. Permanent grassland and leys of varying duration are the source of 'ley pests'; in addition volunteer cereal plants and grass weeds attract the egg-laying females of some pests so that dirty stubbles constitute an important risk in intensive or continuous cereal culture. Frit fly and its allies, common rustic moth, grass moth, leatherjackets, wheat shoot beetle and wheat flea beetle are examples of 'ley pests'.

Methods of cereal husbandry profoundly influence the occurrence and intensity of pest attacks. In recent years the switch to wheat and barley at the expense of oats and rye has meant that stem nematodes have declined sharply in importance. The tendency towards very early sowing (i.e. before mid-October) of autumn cereals has increased the risk of BYDV infection transmitted mainly by the bird-cherry aphid and has rendered crops more liable to attack by frit fly, gout fly and yellow cereal fly; on the other hand, early sowing lessens the risk of damage by field slugs and masks the effects of wheat bulb fly larval damage. Methods of straw disposal and pre-sowing cultivations are important in the context of pest damage; slug attacks are often severe on medium and heavy soils when straw is buried rather than burned and slug damage frequently occurs in direct-drilled crops. Wheat bulb fly females lay more eggs where the soil surface is rough than where the tilth is fine, and more in dry soil than wet. Herbicides which deal effectively with grass weeds and volunteer cereals are an obvious weapon to assist in the cultural control of several pests.

Cereal monoculture involving several successive crops of wheat or barley has not (as yet) led to increasing pest problems; rather the reverse. The expected increase in numbers of cereal cyst nematodes has been prevented, at least partly, by the activities of fungal parasites which occur naturally in British soils.

Damage by cereal pests ranges from total crop loss to inconsequential feeding. Damage assessment methods have been developed for some important cereal pests and economic thresholds have been set above which control measures are cost-effective. There are assessment methods for grain aphid, leatherjackets, frit fly, wheat bulb fly, field slug and cereal cyst nematode. Some cereals, particularly winter wheat, have tremendous powers of recovery from pest attack and can overcome quite extensive damage by such pests as the stem-boring larvae of the yellow cereal fly.

Because of the sporadic nature, both in time and place, of cereal pest attacks in the United Kingdom, prophylactic control measures are often unnecessary and should be reserved for occasions when early warnings of impending damage are issued by Advisory Services. Pest forecasting methods need to be accurate and ideally applicable to individual crops. Techniques based on aerial suction traps (cereal aphids), soil sampling for eggs (wheat bulb fly) or cysts (cereal cyst nematode) or surface-baited traps (field slug) do not necessarily satisfy these criteria and much remains to be done in this direction.

Having established that a particular crop is at risk to a cereal pest, the farmer has recourse to one or more control measures which can be grouped under (a) resistant varieties, (b) cultural control and (c) chemical control. Plant breeders have given relatively little attention to cereal pests, although they have provided oat varieties resistant to stem nematode and barleys resistant to cereal cyst nematode. The partial resistance of certain winter and spring wheat varieties to cereal aphids has yet to be exploited. Cultural control measures are of primary importance; examples are the early destruction of volunteer cereal plants and grass weeds before sowing autumn cereals, adjustment of time and depth of sowing, and adequate use of seedbed and subsequent fertilisers.

Chemical control measures are often used unnecessarily; they should be reserved for situations where other measures are inadequate and then applied with due regard for beneficial organisms, wildlife and the general environment. Insecticidal seed treatments involving minimal dosages of active ingredient are preferable to granule or spray treatments whenever appropriate. Aphicides which are selective in action against cereal aphids are preferred to broad-spectrum aphicides. In practice it is often possible to use two or more different methods to keep pest damage below an economically important level; this is the basis of integrated pest management.

Some important cereal pests

Cereal aphids

Several species of aphids (greenfly) live on cereals. They cause damage directly by sucking sap from plant tissues; barley is more tolerant of such injury than wheat. They may also transmit BYDV from infected to healthy plants. Aphid feeding activities can result in loss of yield and quality of grain.

The bird-cherry aphid *(Rhopalosiphum padi)*. This aphid lays eggs in autumn on bird-cherry *(Prunus padus)*, a common plant in the north of Britain. In more southerly counties the aphid overwinters in the 'summer' form on cereals and grasses. Winged migrants alight on early-sown emerging cereal crops from August to October and may transmit BYDV strains to the seedlings. The progeny of these migrants may infect neighbouring plants unless an aphicidal spray treatment, preferably of a synthetic pyrethroid, is applied in late October or early

November. Bird-cherry aphids living on cereals and grasses within fields may transmit BYDV – a common phenomenon in south-west England and south Wales. Destruction of grass weeds and volunteer cereals with a quick-acting herbicide, followed by suitable cultivations and supplemented by aphicidal spray treatment, may be required in such cases.

The grain aphid *(Sitobion avenae)*. The whole life of the grain aphid is spent on cereals and grasses, overwintering as an egg, or, in mild seasons, in the 'summer' form. It is the reddish-brown aphid frequently seen colonising the stems and ears of flowering cereals. Although the grain aphid is capable of transmitting mild and virulent strains of BYDV in autumn and spring, its chief importance probably lies in its direct feeding activities near the flowering and grain-filling periods. Economic losses in grain yield and a reduction in bread-making quality are often recorded. Aphid parasites and predators play an important role in checking the early spring increase in number of grain aphids. It is therefore advisable to use selective aphicides to deal with infestations at or near flowering.

The rose-grain aphid *(Metopolophium dirhodum)*. This aphid overwinters as an egg on rose. Winged forms migrate in spring to cereals and grasses. Colonies of this aphid are usually confined to the undersurface of cereal leaves. Yield losses are normally less than those attributable to grain aphids, although in the epidemic year of 1979 losses in grain yield amounted to 10%. The rose-grain aphid can act as a vector of BYDV but it is probably of minor importance in this respect. If required, an aphicidal spray treatment at or near flowering gives effective control.

Leatherjackets

These are the legless larvae of several kinds of crane flies ('daddy longlegs') of which the most common is *Tipula paludosa*. Adult flies are found in autumn and lay eggs in grass fields or grass-infected crops. The eggs hatch in about 10 days and the larvae commence feeding on cereals and a range of other crops. Although some damage is noted in autumn and winter, the main feeding period is in the following spring, when spring cereals after grass are particularly susceptible. Cereal plants are chewed off or bitten at or just below the soil surface. Feeding ceases in July and the larvae pupate in the soil. *T. paludosa* has only one generation per year.

Leatherjacket numbers fluctuate widely over the years. Young larvae are extremely susceptible to desiccation, and large numbers of older larvae are eaten by birds. Forecasts of the likelihood of serious damage to cereals are made regionally by entomologists in the Advisory Services. Early ploughing of grass and grassy stubbles to prevent egglaying is a useful cultural method of control. For chemical control several insecticides are available as sprays, granules or bait formulations.

Frit fly *(Oscinella frit)*

This fly is an important pest of grassland and weed grasses as well as cereals, having three generations per year. Larvae of the first generation attack late-drilled spring oats in May and June, causing the centre shoot to turn yellow and die ('deadheart'). Second generation larvae attack oat grains in the panicle, causing blind spikelets or black and shrivelled kernels. Larvae of the third (autumn) generation hatch from eggs laid in grasses, mainly rye-grass, in August and September. The larvae normally overwinter within grass shoots, but when infested grassland is ploughed to cereals, the larvae can migrate to the young cereal seedlings (wheat, barley, oats or rye) and kill the shoot.

Wheat bulb fly *(Delia coarctata)*

Wheat bulb fly is a serious, albeit local, pest of winter wheat in the eastern half of the British Isles. It also attacks winter barley and winter rye but oats are immune. Spring wheat and spring barley sown before early March can suffer serious damage.

Female flies lay their eggs in July and August in bare soil or in soil under crops with open foliage, so that attacks occur in winter wheat or barley following a full or bastard fallow, after root crops including potatoes, sugar beet, onions, carrots or celery, or after early-vining or dry-harvested peas.

Eggs hatch from early January onwards, although the hatching process may be delayed for several weeks in very cold weather. The legless maggots enter the plant and feed within the central shoots, which turn dull, yellow and then die ('deadheart' symptom). Usually in late March or early April, the maggots migrate to other shoots of the same or neighbouring plants, and crop damage often becomes severe at this time. Larvae usually finish feeding in May, leave the plants and pupate in the upper layers of the soil. There is only one generation each year.

Although all commercial varieties of winter wheat are susceptible, those with a free-tillering habit are better able to recover from attack. On fields at risk from the pest, winter wheat should be drilled before mid-October and to a shallow depth. Alternatively, winter oats (which are immune) can be sown, or spring wheat or spring barley if sown after mid-March. Chemical control measures are a useful adjunct to the cultural measures outlined above.

Wireworms

These are the larvae of certain click beetles (chiefly *Agriotes* spp.). The adult beetles are active from April until July and lay eggs in clusters just below the soil surface, chiefly in grassland. Young wireworms hatch in one month and begin feeding immediately. They take four or five years to attain maturity and have two main feeding periods – early March to late May and September to October. Fully fed wireworms pupate in a cell in the soil and the adult beetles emerge in September but remain in the cell until the following spring.

Damage symptoms vary with the size of the plant and the size of the wireworm. Young cereal seedlings are bitten below ground level, become yellow and die. Older plants often show a yellowing of the centre shoot, the outer leaves being unaffected. Attacked tissues often have a characteristic frayed appearance.

Wireworm numbers declined, even in permanent grassland, in Britain from the late 1950s, probably as a result of the widespread use of persistent soil insecticides.

Wireworms are eaten by rooks and other birds and are attacked by certain soil insects and fungi. When ploughing old grass, February or March is preferable to June or July ploughing. Cereals should be sown into a firm seedbed adequately supplied with nutrients. Rolling and top-dressing with nitrogen fertiliser help crops to recover from attack.

Cereal cyst nematode *(Heterodera avenae)*

This species is widespread in Britain and elsewhere and is most damaging in light-textured soils. Oats suffer from attack more severely than other cereals; wheat is more susceptible than barley, while rye is only slightly affected.

Juvenile nematodes invade cereal root systems in large numbers in early spring and to a lesser extent in autumn. Attacked roots are short and much branched, giving a shallow, knotted root system. Some juveniles develop into females which swell, become lemon-shaped and protrude from the root surface; others become males which remain worm-like and emerge from the roots to fertilise the females, which are at first white than turn brown to form cysts that become detached from the roots. Each cyst contains up to 400 eggs, and a proportion of these hatch each year, even in the absence of a host crop.

As the oat acreage has dropped, damage to cereals has become less apparent in the last 20 years. Fungal parasites of eggs and cysts exert natural control of the pest and are chiefly responsible for the decline in numbers of cereal cyst nematode in this country.

A two- or three-year break from cereals, using grass or non-host crops, is often sufficient to bring the pest down to non-damaging levels. Growing barley or rye in place of oats is an obvious cultural method of control. There are barley cultivars which are resistant to one or more races of the nematode. Unfortunately, resistant oat cultivars are sensitive to invasion and do not yield well in infested soils. Chemical control measures have been used successfully in experimental work, but are far too costly for general adoption.

Stem nematode *(Ditylenchus dipsaci)*

The stem nematode is common and widespread, attacking oats and rye to cause 'tulip root' or 'segging'. Wheat and barley are not affected in this country. Weeds such as cleavers, mouse-ear chickweed and wild-oat are susceptible and may carry-over the pest from one oat or rye crop to the next.

Male and female nematodes are plentiful in the stem and leaf tissues of attacked plants. Eggs hatch within a few days and juveniles mature in about four weeks.

Control is extremely difficult because the race/s which affect oats and rye also attack many other crop plants and several weed species.

Varieties of oats which are resistant to stem nematode attack are now available. Chemical control methods have not been developed for oats or rye.

Field slug *(Deroceras reticulatum)*

Although several species of slugs can attack cereals, the one which is of paramount importance is the field slug. It attacks autumn-sown cereals in medium to heavy moisture-retentive soils after grass, clover, oilseed rape, maize, peas and beans. Winter wheat is particularly susceptible, but spring-sown cereals can also be damaged. Late-sown winter wheat is more at risk than wheat sown before mid-October.

Slug damage starts within a few days of drilling, when the germ is removed from the seed below ground. Young seedlings may be grazed at ground level and their growing points are severed. Shredding of the leaves above ground is often observed but this is of less consequence than the subterranean earlier damage.

Birds and mammals

A number of birds, including protected species, may be responsible for causing damage to cereals by removing the seed, pecking the newly emerged shoots, pulling up whole seedlings, or feeding on the ears just before harvest. Scaring devices are only partly effective against bird damage. Licences can be issued under the Wildlife and Countryside Act 1981 to enable farmers to shoot protected species as an additional measure when crop loss potential is high.

Newly sown cereals may be dug up by rats and field mice, but the main damage by mammals is the grazing in late autumn, winter and spring by rabbits, field mice, hares, deer and voles. Towards harvest, rabbits, hares, rats and grey squirrels may bite through cereal stems. Shooting, trapping and gassing are measures used to prevent or reduce mammal damage.

Pest identification

Assistance with identification of cereal aphids may be obtained from the Pest Identification Card series (coded IC/90 to IC/95) which are available from MAFF Publications, Ministry of Agriculture Fisheries and Food, Alnwick.

Chapter 3 HUSBANDRY AS AN AID TO INTEGRATED CONTROL

Many aspects of cereal husbandry affect the incidence of pests, diseases and weeds in the crop. Some have major effects and can be regarded as important control measures; in some cases a change in husbandry practice may even lead to the virtual disappearance of a particular pest, disease or weed. In other cases, effects may be much smaller but it is important that they are recognised because a small adverse effect may make an existing problem worse or, conversely, changing a technique which slightly favours a particular pest, disease or weed may play a significant part in an overall control programme. Most important, however, is an awareness of the full implications of the various aspects of husbandry on pest, disease and weed incidence. This, together with a knowledge of chemical control methods, will enable a farmer or adviser to construct a fully integrated system of control.

Soil factors

Soil type. The most significant effect of soil type on pest, disease and weed incidence in modern cereal production systems is indirect; it has a controlling effect on the range of crops which can be grown and hence on crop rotation. It will also have a marked influence on a farmer's choice of cultivation methods. Both these things have considerable implications for pest, disease and weed control and are discussed later in this chapter.

Soil type and climate together directly determine the potential weed flora on a particular site; examples are given in all textbooks on botany. The direct influence of soil type on pests and diseases is less marked but cereal cyst nematode *(Heterodera avenae)* is a typical example of a pest which rarely causes trouble on the heavier loams and clays but may be a serious problem on light land. On the other hand, slugs are rarely found in quantity on sandy soils but are much more common on silts and clays. Soil type may also affect the severity of attack by soil-borne diseases; take-all *(Gaeumannomyces graminis)*, for instance, is normally much more pronounced in lighter soils which not only favour infection but also, because of their proneness to drought, accentuate the effects of the root loss caused by the disease.

A further consideration is the effect of soil type on the efficiency of soil-acting herbicides. In organic soils (usually taken as soils with an organic matter content above 10%) herbicides of this type may be inactivated because of absorpiton on to the organic matter. In light sandy soils these herbicides leach more rapidly from the soil surface into the rooting zone of the crop, reducing the chances of effective weed control and increasing the risk of crop damage. The extent of these effects varies widely with the physical and chemical characteristics of the individual herbicides. Similar problems may arise with soil-acting pesticides and fungicides.

Drainage and soil structure. Soil which is waterlogged for any appreciable time produces a characteristic weed flora which may be extremely difficult to control until the problem is cured. Cereals growing in very wet soil lack vigour and this aggravates the effects of soil-borne pests and diseases. The risk of crop damage from soil-acting herbicides also increases. Waterlogging normally occurs because either the underdrainage is ineffective or faulty soil structure is preventing excess water from moving away from the rooting zone.

Soil acidity and liming. Acid soils too produce a typical weed flora and the addition of lime is the first essential step in its control. It should be noted that some weeds such as corn spurrey, *Spergula arvensis*, which tolerate acid conditions, grow more strongly at a higher pH. Cereals do not grow well in acid conditions, although oats and rye are more tolerant than wheat and particularly barley. The poor crops normally found on acid soils are very prone to damage by soil-borne pests and disease. Take-all is anomalous in that acidity often makes plants more susceptible to attack, but fungal growth is favoured by alkaline conditions. In some cases the disease has been particularly severe in the year after lime has been applied to correct acidity. Soil pH may affect the activity of some residual herbicides.

Plant nutrient supply. Although a marked deficiency of either phosphate or potash is seen only rarely nowadays in farmed land, typical weed flora may develop if a deficiency exists. Crops suffering from a serious shortage of phosphate, potash, magnesium or one of the trace elements compete only poorly with weeds. Because growth is already restricted by the deficiency, they are more susceptible to damage by soil-borne pests and diseases. Phosphate is important to combat wireworm *(Agriotes* spp.*)* and gout fly *(Chlorops pumilionis)* damage. Take-all too is greatly favoured by phosphate deficiency.

An inadequate supply of nitrogen reduces the competitiveness and vigour of a cereal crop and often accentuates pest or soil-borne disease problems. However, the response to top-dressed nitrogen is usually quick. Because of the flexibility in yield structuring in cereals (particularly in wheat), it often helps a crop to recover more rapidly and more fully from an attack of a pest which has reduced plant or tiller number such as wireworm, wheat bulb fly *(Delia coarctata)* or frit fly *(Oscinella frit)*. Extra spring nitrogen also helps to overcome the effects of root loss caused by take-all. Although nitrogen does not affect the incidence of a pest or soil-borne disease or the level of damage, it alleviates the effects of the damage and so it may properly be regarded as an important item in an integrated control programme.

Nitrogen may affect pest, disease and weed problems in other ways also. Dealing first with weeds, the responses of the cereal crop and the weed species to added nitrogen are usually different. In most cases this is to the advantage of the farmer because the crop shows the better response. Black-grass *(Alopecurus myosuroides)* in winter wheat provides a typical example where the competitive effect of the weed is reduced if adequate nitrogen is given. With wild-oats *(Avena* spp.*)*, however, there is evidence in some cases of a better response to nitrogen than from the cultivated cereal and although this is rare, nitrogenous manuring cannot be regarded as a useful control measure. Similarly chickweed *(Stellaria media)*, particularly under cold, wet conditions, may respond to nitrogen more readily than the cereal crop and become aggressive.

In the case of foliar disease, high levels of nitrogen usually increase the severity of an attack. This is well documented with mildew *(Erysiphe graminis)* and rusts *(Puccinia* spp.*)* and there is evidence of similar effects with other diseases. The increase is usually recorded over the range of nitrogen levels leading up to as well as beyond the optimum rate. It has been shown that the efficiency of nitrogen utilisation is often improved if foliar and ear diseases are controlled and a farmer should be aware of the particular need for fungicide protection where nitrogen use is generous. There is also the possibility that grain aphids *(Sitobion avenae)* may multiply more rapidly on crops of high nitrogen status.

Variety

The principal factor in choice of variety in the present context is resistance or susceptibility to pests and diseases. There are relatively few examples of resistance to pests, but varieties of spring barley and both winter and spring oats with resistance to cereal cyst nematode are now available. So, too, are varieties of winter oats resistant to stem nematode *(Ditylenchus dipsaci)*. These pests are difficult to control in other ways and stem nematode, with a range of alternative hosts, may be a particular problem on some soils in some cropping sequences. Clearly, the choice of a resistant variety should be standard policy on land known to be infested.

Turning to diseases, many current varieties of winter wheat possess a reasonable degree of genetic resistance to eyespot *(Pseudocercosporella herpotrichoides)* and it is sensible to use those with the best ratings in intensive cereal systems where the risk of the disease is greatest. With foliar diseases the position is more complicated. Very few varieties of any cereal species have good resistance to all the common diseases. Most farmers should therefore choose several varieties with different patterns of resistance to grow on the same farm to reduce risks from an attack of any particular disease. In some areas farmers may be better advised to choose varieties with good resistance to a disease particularly prevalent in their district rather than a range of patterns of resistance.

With mildew in spring barley and mildew and yellow rust in winter wheat, further discrimination in choice of variety is possible because the different resistance factors to these diseases have been identified and varieties grouped according to the factor which each possesses. Risks of serious attacks of either disease can be minimised by growing varieties from different groups. Details are given in NIAB Farmers' Leaflet No 8 *Recommended Varieties of Cereals*, the National Institute of Agricultural Botany, Cambridge, which is revised annually. An extension of this approach is to grow a crop made up of a mixture of varieties with different factors of resistance to the main disease threats. This has not proved a very popular technique and can pose problems in marketing for milling, malting or export when purchases may be based on the varietal purity of the load.

The susceptibility of some cereal varieties to damage by certain crop protection chemicals is an additional point to remember when planning an integrated control programme. It is found most frequently with herbicides (for instance, difenzoquat and chlortoluron on some winter cereals) but may also arise with insecticides.

Occasionally other varietal characteristics may be of significance. Many years ago the winter wheat variety Jubileegem was highly regarded in some heavy land areas because its profuse tillering helped to suppress black-grass. More recently the relatively poor vigour of Maris Huntsman in winter and early spring may have aggravated weed problems.

Crop rotation

Crop rotation is one of the most potent cultural methods of pest, disease and weed control. Historically, the diversity of crops needed to provide an adequate rotation, and the different cultivation which they needed at different times of the year prevented a build-up of pests, diseases or weeds specific to or closely associated with a particular crop. Sequences likely to lead to problems were avoided. Some of the classical features of rotation such as the inclusion of crops which needed or allowed inter-row cultivation for weed control are rarely exploited nowadays. Instead, a new advantage is that the well-planned use of herbicides over a rotation gives flexibility in weed control as a result of the wider range of herbicides available for use in a mixed cropping sequence than in monoculture.

The availability of crop protection chemicals has removed the absolute necessity for crop rotation in many cases, but the advantages are very considerable and they should not be ignored lightly. Again it is a question of the overall profitability of the farming system. In the past it was often necessary to grow crops which provided more diversity in the rotation but were themselves unprofitable. They were tolerated because they improved the profitability of the farm as a whole. Partly because of changed economic circumstances but principally as a result of the wide range of crop protection chemicals available, a farmer has more difficult decisions to make nowadays. If he wishes to alter his farming system by omitting a relatively unprofitable crop, he must decide whether the advantages will outweigh the cost of more agrochemicals or other control methods which may be needed as a result of the change. (He must, of course, also assess the implications for such items as fertilizer and labour requirements which are outside the scope of the present discussion.) It is not a simple exercise and the role of rotation in integrated control systems can best be shown by outlining the principal types of rotation in which cereals are grown.

Ley arable rotations. These are usually found nowadays in the wetter areas of the country. The arable crops grown normally include fodder roots and potatoes on the lighter soils while cereals are likely to predominate on heavy land. In the first sub-group, pest, disease and weed problems in cereals are minimal. The continually changing cropping on any one field prevents the build-up of soil-borne pests and diseases specific to cereals and of any appreciable weed problem. The main risk will be to cereal crops following immediately after grass. Problems may include frit fly especially if the ley contained rye-grass and other 'ley pests' as discussed on p. 36; annual grasses if these have been allowed to shed seed, and perennial grass weeds if present in the ley. Often non-specific soil-inhabiting pests such as leatherjackets *(Tipula* spp.) and wireworms *(Agriotes* spp.) may be more of a problem. Crop damage by these species is often compounded by the difficulty of preparing a fine, firm seedbed after a ley, which would encourage early root development and growth and so speed recovery from attack.

On heavier soils where the arable part of the rotation consists mainly of cereals, the same problems can be expected after the leys and there may be others too. Particularly if the emphasis is on autumn-sown cereals, grass weeds often reach significant levels. In the relatively high rainfall areas where rotations of this type are frequently found, harvest is usually late and opportunities for autumn cleaning are very limited. This may aggravate problems with perennial grasses such as rough meadow-grass *(Poa trivialis)* and rye-grass *(Lolium perenne)* which re-established from seed as well as surviving as transplants. The same circumstances lead to a high risk of slug damage, eyespot and possibly take-all. All these problems can be reduced by including spring-sown cereals in the rotation.

Mixed arable rotations. Mixed arable cropping is widely practised on the deeper light-medium upland soils and on fen peats and the lighter silts. As well as cereals, crops grown may include cash and forage roots, pulses, field vegetables and one-year leys. As with a ley/arable system which includes a good diversity of cropping, no serious problems with weeds or soil-borne diseases are likely to arise. The most common pests are wheat bulb fly and, on the heavier soils in the group, slugs. Wheat bulb fly is seen typically after early potatoes or vining peas; the adults lay their eggs on bare ground in July and August after these crops have been harvested. Slugs may give trouble in wet seasons after leafy crops such as peas, beans and occasionally sugar-beet which provide good conditions for the field slug *(Deroceras reticulatum)* to multiply. It is this species which causes most damage to the germinating seed and young plants of cereals; the garden and keeled slugs which often damage potatoes rarely harm a following cereal crop.

Cereals with combinable break crops. These systems first developed on the heavier soils in the drier parts of the country and rapidly spread to some of the thinner chalk and limestone soils. They are now widespread. The breaks from cereals are provided by oilseed rape and peas or beans for grain production. Typically, at least three-quarters of a farm is cropped with cereals and the proportion is often higher. Most farmers prefer to sow all the crops in the autumn because of the higher yields obtained.

In these situations the small proportion of break crops provides little or no control of soil-borne pests and disease of cereals; it sometimes happens that including rape or peas increases slug problems. Because of the urgency to complete sowing by mid-October, cereal sowing usually starts in September, increasing the risk of infection with BYDV and of damage by several insect pests, as well as the chances of infection with leaf diseases from volunteers from the previous crop. However, the break crops do provide an opportunity to use different herbicides and in particular materials which are very active against a wide range of grass weeds. This is helpful in the control of wild-oats and black-grass and particularly useful in dealing with species such as the brome grasses *(Bromus* spp.*)* which are difficult to kill in cereals. The relatively early harvesting of rape (and of winter barley, if it is grown) provides a longer period for weed control by cultivation than between two winter wheat crops.

Cereal monoculture. Continuous spring barley was the first form of cereal monoculture to be seen in this country, mainly on thin chalk and limestone soils, but since about 1970 continuous winter wheat has been grown successfully on many of the heavy soils in East Anglia. More recently, winter cereals have been replacing spring barley on the chalks and limestones. It might appear that systems of this type provide no rotational control at all but there is one unusual example. The take-all fungus can be found in all fields growing continuous wheat or barley but has only small effects on yield unless crop growth has been checked by other factors such as poor root development or drought. This is due to the phenomenon known as take-all decline which reduces the severity of the disease. It is probable that take-all decline also occurs in intensive cereal systems which are not monocultures. Experiments have shown that a one-year break from white straw crops may not lead to the complete disappearance of the effect. Break crops appear to vary in this respect, some destroying take-all decline more effectively than others.

The only other way in which cropping sequence may help pest, disease or weed control in cereal monoculture is if a spring-sown crop is introduced in a series of autumn-sown crops. This may assist in dealing with serious weed problems, but the lower yield from the spring-sown cereal, particularly on heavy land, is usually unacceptable.

In spite of these two features, it must be recognised that a farmer practising cereal monoculture has foregone most of the benefits normally associated with rotational cropping in terms of disease and weed control. If he also chooses to sow all his crops in the autumn, very few alternative methods of cultural control are available and heavy dependence on agrochemicals must be accepted, together with inevitable problems such as soil-borne diseases which cannot be controlled chemically. Cereal monoculture does not appear to have aggravated pest problems.

Method of straw disposal

The method of straw disposal is usually decided by factors other than pest, disease or weed control. If a farmer needs straw for his own livestock or has a good market available, it will be baled and removed from the field. Until recently, in most other cases the straw was burned. The practice of chopping straw and then incorporating it into the soil has been slow to find favour in the UK. It is usually unnecessary in order to maintain soil organic matter levels and can be harmful to autumn-sown cereals in the establishment phase. However, incorporation is becoming more common to avoid the environmental problems which have arisen from burning straw.

The method of straw disposal normally has little effect on pest, disease or weed incidence if the subsequent crop is not sown until the spring but there are considerable effects in sequences of autumn-sown cereals. If the straw has not been burned, weed problems are usually worse, particularly with annual grasses such as black-grass and barren brome *(Bromus sterilis)*. Poor incorporation of the residues of the previous cereal crop reduces the effectiveness of many soil-acting herbicides and also increases the risk from diseases spread by rain-splash such as net blotch *(Pyrenophora teres)* in barley. Again, the presence of trash on or near the soil surface exacerbates slug problems, particularly in wet autumns.

Cultivation

Cultivation is a valuable method of assisting the control of many weeds and some pests and diseases. Not many farmers nowadays choose a cultivation system primarily for these reasons but in some cases effects on pest, disease and weed incidence are significant and in all cases it is important that the effects of cultivations are understood.

Stubble cleaning. Stubble cleaning cultivations are seen much more rarely now than in the past. This is due in part to the increasing trend to intensive cropping with autumn cereals and in part to the availability of cheap and effective chemical alternatives. However, the need for stubble hygiene is particularly great in this type of cropping sequence and it is most important that the opportunity to reduce weed populations and the risks of pest and disease carry-over on cereal volunteers or grass weeds should not be neglected.

Preparation for sowing. Although there was a considerable move away from ploughing in the 15-20 years up to 1980, it remained the most usual primary cultivation for cereals, especially for spring sowings. It is mainly on the heavier soils in the drier areas that the swing towards reduced or zero cultivation was most marked. With successive crops of winter cereals becoming standard practice in many cases, non-ploughing methods were particularly attractive. The term "reduced cultivation" has been interpreted by farmers in many different ways but the main objective has been to avoid deep working and the production of clods, so that the speed of primary cultivation is increased and the number of secondary cultivations reduced. The better structured soils are particularly suited to shallow cultivation techniques and in some cases direct drilling has proved an effective policy. However, a realisation of the value of ploughing in controlling pests, diseases and weeds, coupled with improvements in design, has now led to more widespread use of the plough again, even on relatively difficult soils.

The main advantage of mouldboard ploughing in the context of weed control is that it buries weeds and weed seeds. Annual weeds rarely re-emerge after burial to plough depth and many weed seeds either rot or germinate but fail to emerge. However, seeds of some species, notably wild-oats, may lie dormant in the soil for a number of years and so maintain a stock of seed in the land. In soil which has been direct drilled or cultivated only shallowly, weed seeds remain close to the surface. Although temporary dormancy may be induced by light or temperature effects in some species, most of the seeds will germinate rapidly in their normal season. Because many are on the surface, some may be removed by birds and small mammals and others rot and decay. On balance, ploughing reduces annual weed populations to a useful extent but dormancy of buried seed encourages the perpetuation of a few species. Shallow cultivation or direct drilling may give high weed populations at first but these can be reduced rapidly by effective herbicides because no store of seeds is being incorporated into the soil. There is, however, a greater risk of weeds reaching economically significant levels if herbicide treatment is neglected or badly judged. Between the two extremes of mouldboard ploughing and direct drilling there lies a whole range of cultivation techniques, their effects on annual weeds depending on the degree of soil disturbance and inversion in each case.

Pre-planting cultivations affect trash burial and soil aggregate size and may therefore influence herbicide activity. If trash remains on the surface, or if there is a layer of carbon residue following straw burning, the effect of soil-acting herbicides may be considerably reduced. A very cloddy seedbed may lead to poor results too. Parts of the soil surface are sheltered by the clods from the spray mist and receive an inadequate dose, while soil in the centre of a clod which becomes exposed after weathering often contains weed seeds but no herbicide.

Cultivation method is a significant factor in the control of perennial weeds. With rhizomatous species such as common couch grass *(Elymus repens)*, at least on mineral soils, a plough can usually be set to undercut the layer of soil containing the rhizomes. Once this has been done, a tined cultivator can be used either to stir and move the disturbed layer so that it dries out or to pull the rhizomes to the surface where dehydration will be more rapid. When the time interval and weather permit, successive cultivations can be used to kill regrowth from rhizome fragments and exhaust the food reserves in them. If a plough is not used initially, it is much more difficult to separate the rhizomes from the lower layers of the soil. Shallow cultivation should be avoided with weeds of this type because it tends to spread broken pieces of rhizome without uprooting the main infestation. The same may happen with stoloniferous species.

Deep rooting perennial weeds such as coltsfoot *(Tussilago farfara)* and field bindweed *(Convolvulus arvensis)* are more difficult to control by cultivation. However, ploughing is normally preferable because it severs the top portion of the root completely, making it necessary for the main rootstock to devote more of its reserves to the production of fresh growth. It will then be rather more susceptible to subsequent chemical control measures.

Pre-planting cultivations may also affect the severity of damage by soil-borne pests and diseases. It is important that no cultivation should produce a consolidated layer or pan in the soil which would restrict root growth and so exacerbate attacks of root diseases such as take-all. Cereals are more likely to survive and recover from attacks of such pests as slugs, wireworms and leatherjackets if the seedbed is fine and firm. A firm seedbed also reduces the risks of severe take-all; the fungus runs very freely over roots in a “puffy” soil. Particularly on heavy land, a fine, firm tilth is often difficult to prepare after ploughing or deep tined cultivation which frequently produce large clods. On the other hand, the risk of infection with some fungal diseases in successive cereal cropping can be minimised if all residues of the previous crop are effectively buried by ploughing.

Post-planting cultivations. Cultivation during the tillering period in the spring was once a common method of weed control in cereals. Inter-row hoeing disappeared with the horse but harrowing is still sometimes seen. In the right conditions with small weeds, a vigorous crop and good drying weather, it can control annual weeds effectively with very little crop damage. However, harrowing is difficult to fit into an integrated control programme. There are restrictions on its use in crops which have been or shortly will be sprayed with a number of crop protection chemicals.

Although similar restrictions affect rolling, this operation remains a valuable method of helping a cereal crop to recover from an attack by one of the soil-inhabiting pests or of take-all. The consolidation of loose "puffy" soil provides better rooting conditions which are quickly reflected in top growth. Unlike harrowing, which rarely gives better weed control than one of the cheaper herbicides, there is no substitute for rolling in encouraging a cereal crop to grow away from root damage and consequential limitations on the use of agrochemicals must be accepted.

Time of sowing

Delayed sowing is a time-honoured method of controlling annual weeds. Typically, a rough seedbed is prepared and left undisturbed for as long as possible before the weeds are killed either by cultivation or a non-selective herbicide. This technique survived at least as late as 1970 for controlling black-grass in winter wheat on heavy land. As with some other cultural control methods, its success depends largely on weather conditions which affect both the degree of weed control achieved and the growth of the late-sown crop. It is usually unwise to delay drilling autumn-sown crops beyond the end of October because of the risk of yield depression. Although sowings made in November or even later give satisfactory results in some years, the chances of success are reduced. In the spring, dates are less meaningful because of soil differences and geographical and seasonal variations in weather. However, to delay beyond the time when soil moisture and temperature are suitable for sowing involves the risk of yield loss because it reduces the length of the growing season. In general, there are now few cases in either autumn or spring where delaying sowing beyond the normal time solely for weed control purposes can be justified.

On the other hand, the manipulation of sowing date should not be ignored as a significant method of affecting pest, disease and weed incidence. In recent years sowing winter cereals very early in the autumn (before the last week in September) has become increasingly popular. There is evidence that mid-September sowing is an advantage on some of the lighter soils and in the North, where early establishment is a necessary foundation for high yield. In other cases there has been no advantage in sowing before late September and rarely any yield loss from sowings made up to mid-October in the main cereal growing areas of England.

In all cases, September sowing entails much greater risks of problems with pests, diseases and weeds. Of the pests, the yellow cereal fly *(Opomyza florum)* and the bird-cherry aphid *(Rhopalosiphum padi)*, the main vector of BYDV, are rarely found except in September-sown crops; attacks of gout fly and frit fly are usually more serious when the crop has been sown very early (in contrast to the position with spring-sown oats, where earliness is essential to reduce the effects of frit fly). However, damage from wheat bulb fly is usually less serious on sowings made in September. Because opportunities for disposing of trash and volunteers are reduced and because crops emerge when temperatures are still high enough to favour disease spread, very early sowing favours a number of cereal diseases including mildew, yellow rust, net blotch and eyespot. The incidence of take-all is higher in early sown crops too. There is also a risk of worse weed problems. With a shorter interval before sowing, fewer weed seeds can germinate and then be killed before the crop is sown. Very effective herbicide treatment is therefore necessary if competition with the crop is to be avoided. In addition, early sowing leaves a longer period for weeds to germinate after the application of a herbicide. Currently available materials may not be persistent enough to give season-long control and repeat doses are often needed.

Other cultural factors

High plant populations rarely reduce yield in cereals if steps are taken to prevent serious disease levels or lodging from developing. An above average seedrate is therefore a sensible precaution when an attack by pests which may kill the plants or tillers early in the life of the crop is expected. Although there is some evidence that seedrate and spatial arrangement of the plants may affect weed competition, these are unlikely to be significant factors with the husbandry practices commonly used in the UK.

Depth of drilling may be important in some cases. In general, shallow sowing favours quick establishment and rapid early growth which will help the crop to survive attacks from pests such as wheat bulb fly, to compete more effectively with weeds and may reduce the incidence of diseases such as eyespot. A sowing depth of 20-25 mm is usually enough, unless the surface soil is dry and cloddy or herbicides which require more soil cover over the seed are to be used.

The Significance of Cultural Methods of Weed, Disease and Pest Control

The techniques described above show a considerable diversity both in themselves and in the way in which they affect pests, diseases and weeds. For convenience, each method has been dealt with in turn. Their roles in modern cereal production can be demonstrated more clearly by classifying them according to their main functions: see Table 3.1; the lists are not intended to be comprehensive but cover only the more important methods.

The influences of some of the various husbandry practices available for weed, disease and pest control in cereals are tabulated at Table 3.2 (p. 56) to make it easier to evaluate the likely overall effect of a change in practice.

TABLE 3.1 Some husbandry factors classified by main function

1. **Reducing pest, disease and weed incidence.** Rotational cropping, avoiding very early autumn sowing and maintaining satisfactory soil pH, drainage and plant nutrient status provide good examples.
2. **Reducing existing pest, disease and weed populations.** Weed control by soil inversion, delayed sowing and the adoption of a wider crop rotation fall into this group.
3. **Reducing effects of pest and disease attack.** Choice of variety is important. Crop recovery can be aided by suitable cultivation and by maintaining satisfactory soil water, pH and plant nutrient status.

Alleviating the effects of pest or disease attack is a unique function of cultural control methods and it is difficult to envisage an agrochemical filling the same role. In contrast, chemicals provide effective alternatives for reducing populations of most pests, diseases and weeds although suitable materials are not always available. In the case of avoiding pest, disease or weed incidence the position is rather more complicated. Agrochemicals provide a true alternative to cultural methods in only a very limited number of cases but the need to adopt an avoidance policy is less stringent if effective sprays are available to control problems when they develop.

It follows that the role of crop husbandry in reducing the effects of pests, diseases and weeds remains important in modern systems of cereal production although it is less vital than in the past because of the availability of agrochemicals to reduce damage levels. The choice of cultural or chemical methods or a combination of the two for reducing existing pest, disease and weed populations should be based on their relative cost-effectiveness and reliability. This is not always easy to decide because many of the cultural methods also affect other aspects of crop growth and production in ways not directly related to pests, diseases or weeds. Cultural control is, of course, the only option in cases where no effective agrochemical is available.

The choice between pest, disease and weed avoidance by cultural methods and the acceptance of higher levels of incidence requiring higher chemical usage is the most difficult to make. The maintenance of good soil conditions in terms of pH, water relationships and plant nutrient supply is desirable from all points of view. The other potent method of avoidance is by crop rotation and changes in diversity or sequence of cropping may have considerable effects on the whole farming system and hence on farm income. It is particularly important that farmers and advisers should be aware of the increased pest, disease and weed problems which will arise if a traditional type of rotation is abandoned and intensive cereal production adopted. In practice, a change of this type cannot easily be reversed because of the high cost of re-equipping the farm for non-cereal crops. Nevertheless, some flexibility usually remains for adjusting cropping between cereals and the limited area of combinable break crops grown and this should be used either to reduce pests, diseases and weeds generally or to combat a particularly difficult problem.

TABLE 3.2 Influences of husbandry practices on weed, disease and pest control in cereals

Rotation or cropping sequence

Weeds	Diseases	Pests
Continuous winter cropping encourages some difficult weeds.	Intensive cereal cropping increases take-all and eyespot risks, but continuous cereal cropping leads to take-all decline.	Continuous cereal cropping with grass weeds present increases frit fly, cereal cyst nematode, leatherjacket, swift moth caterpillars, field slug and cereal aphid (BYDV). Clean stubbles reduce pest problems.
Continuous spring sowing generally reduces problem weeds.	Repeated cropping with the same cereal species favours trash borne leaf diseases. (*Septoria*, *Rhynchosporium* and net blotch.)	Cereals after grass, especially if interval between ploughing and sowing less than 6 weeks, increases risk of damage from wireworms, frit fly and related species, leatherjackets, caterpillars of grass moths and swift moths, cereal aphids (BYDV) and field slug.
Break crops allow use of some very effective grass weed herbicides.	Ploughing before a break and not ploughing after it or *vice versa* increases the effectiveness of a one year break in controlling eye-spot.	Cereal after full or bastard fallow, root crops or vining peas increases wheat bulb fly damage risk in wheat and barley (oats immune).
Grass break may introduce grass weed problem.		
Crops may persist as weeds e.g. winter barley in winter wheat.		

Soil fertility and fertiliser practice

Weeds	Diseases	Pests
Low weed populations are more competitive in conditions of low fertility.	Take-all aggravated by low phosphate and to a lesser extent by low potash.	Effect of wireworms, gout fly and most soil pests increased by phosphate deficiency and nitrogen deficiency.
	Adequate spring nitrogen enables plants to replace roots damaged by take-all.	Adequate fertiliser assists plants' recovery from pest attack.
High fertiliser use can produce greater response in weeds (e.g. wild-oat) than in cereal crop.	High nitrogen use increases susceptibility to leaf diseases especially rusts and mildew, also increases risk of eyespot lodging.	High nitrogen use increases risk of cereal aphids.
	Manganese deficient plants are more susceptible to mildew.	

Soil conditions, cultivations and straw disposal (in continuous cereal sequences)

Weeds	Diseases	Pests
Direct drilling encourages perennial weeds. Minimal tine cultivations encourage annual grass weeds. Ploughing buries many seeds to depths from which they cannot emerge or even survive (e.g. barren brome).	Direct drilling and minimal cultivations increase the risk of early attack by trash-borne leaf diseases and ergot (wheat and barley). Direct drilling tends to reduce eyespot incidence. Tine cultivations spread barley yellow mosaic virus fungal vector.	Direct drilling and to a lesser extent minimal cultivations increase field slug and frit fly problems when cereal establishment is slow but decrease some stem borers and allow aphid predators to increase in numbers.
Straw burning reduces the amount of weed seed returned to the soil; may encourage the germination of some weed seeds near the soil surface; ash on surface affects some herbicide activity.	Straw burning reduces the inoculum of trash-borne leaf diseases on non-ploughed land. (But very small amounts of inoculum remaining will be sufficient to initiate an epidemic if conditions favour disease development.)	Straw burning which leaves a clean field reduces risk of attacks by frit fly and related species.
Spreading and incorporating straw disperses some weed seeds and may affect some herbicide activity if not deeply mixed or ploughed under.	Poor drainage and poor soil structure aggravate effects of take-all and brown foot rot.	Marked soil consolidation reduces the effects of slugs and wireworms.
		Clean stubbles reduce pest problems.

Seedbed conditions

Weeds	Diseases	Pests
A fine tilth may encourage greater weed germination.	A loose 'puffy' seedbed favours take-all development.	A fine tilth favours egg-laying by wheat bulb fly but decreases field slug activity. A rough cloddy tilth increases field slug but is less suitable for egg laying by wheat bulb fly.

TABLE 3.2 *Continued*

Weeds	Diseases	Pests
Seed rate and plant population		
Low plant populations are less competitive against weeds that survive.	High seed rate may aggravate take-all effect by restricting the root system of individual plants.	High seed rates help to compensate for losses at seed and seedling stages resulting from attacks by birds, slugs and other pests.
Time of sowing		
Early autumn sowing, especially in non-plough situations, increases the risk of annual grass weed problems and broad leaved weeds such as chickweed, speedwell, mayweeds.	In the autumn early sowing can increase the incidence of yellow rust, eyespot, net blotch of barley and autumn mildew. Take-all can also be aggravated.	Drilling before mid-October increase the effect of frit fly, cereal yellow fly, gout fly and cereal aphids (which may carry barley yellow dwarf virus); wheat bulb fly importance will be decreased provided previous crop stubbles remain uncultivated until the end of egglaying.
Delayed drilling in spring and autumn may allow some weed germination and destruction before sowing crop.		Drilling after the end of October increases the importance of wheat bulb fly but decreases the other pests listed above.
Drilling spring cereals before March increases risk of black grass.		In the spring sowing spring wheat before mid-March increases risk of wheat bulb fly. Sowing spring oats before early April decreases frit fly risk; sowing spring barley after early April increases gout fly risk.
Depth of sowing		
Deep drilling leading to delayed emergence may enable weeds germinating near the surface to become more competitive in relation to the crop.		Deep drilling in both autumn and spring increases the effect of most soil pests including wheat bulb fly and minimises the value of insecticidal seed treatments. Shallow drilling and quick establishment of autumn-sown wheat and barley decrease the risk of wheat bulb fly.

Choice of cereal species

Weeds	Diseases	Pests
Oats may limit control measures against wild-oat. 'Open' cereal crops such as wheat and triticale are more prone to competition from spring germinating weeds than winter barley.	Oats suffer greater damage than other cereals from BYDV Take-all has greater effect on winter wheat than winter barley; rye and triticale are less susceptible to infection and oats are virtually immune to the common strains of the fungus; spring-sown cereals usually escape severe attack.	Susceptible varieties of winter- and spring-sown oats suffer greater damage from cereal cyst nematode than other cereal species. All spring-sown oat varieties susceptible to frit fly damage in summer. Spring-sown wheat is more susceptible to most pests than spring-sown barley. Winter-sown oats are immune to wheat bulb fly and resistant to gout fly. Spring-sown oats immune to wheat bulb fly and wheat blossom midge.

Choice of cereal variety

Weeds	Diseases	Pests
Shorter-strawed varieties less competitive against tall weeds. Varietal tolerance to some herbicides varies.	Short-strawed varieties more prone to splash borne diseases such as septoria (wheat) and rhynchosporium (barley). Varieties vary in resistance to leaf diseases and to eyespot. No variety shows good resistance to take-all, but some appear better able to tolerate the depredations of the disease.	Cereal aphids – some wheat varieties partly resistant. Cereal cyst nematode – resistant spring barley and winter oat varieties available.

Chapter 4 AN INTRODUCTION TO CHEMICAL CROP PROTECTION IN CEREALS

The Development of Chemical Crop protection

For most of the agricultural history of mankind, weeds and pests have been kept in check by crop rotations, ploughing, cultivations, hoeing and hand-work, and many of these methods are still of value today. However, even the Ancient Greeks were aware of the time-consuming inefficiency of manual methods of controlling weeds, diseases and pests. The 5th century BC Greek Democritus advised that woody weeds could be cleared by wetting the roots with hemlock juice in which lupin flowers had been soaked. Mixing residual oils from olive oil manufacture with earth to make sterile threshing floors was a practice known to 1st century BC Romans, the mixture probably containing glycosides, glycerides and sea salt as the biotic factors. In the 13th century AD, Arab sources described the use of sea salt as a herbicide, which became used as a selective weedkiller in 16th century England. A 17th century Scottish writer described using salt at 10-12 bushels/acre on wheat, but he also mentioned the difficulty in ensuring adequate crop selectivity. The development of lime-based fungicides and insecticides commenced in the 19th century, and the addition of copper sulphate to form effective fungicides, such as Bordeaux mixture, heralded the start of the modern agrochemical industry and the widespread use of crop chemicals. The discovery of DDT in 1939 and the growth-regulating herbicides in the early 1940s, stimulated the post-war search for new crop protection chemicals.

Recognition by farmers that such chemicals both reduced the need for increasingly expensive labour, and overcame problems that could not be dealt with manually, increased interest and the willingness to pay for such materials. The advent of machinery which could apply such chemicals readily and rapidly to large areas of ground fostered the interest and this, along with the post-war boom in agriculture, led to major agrochemical research and development efforts by chemical companies.

As described in earlier chapters, a competent grower will minimise the risk of weed infestation, disease and pest attack by using cultural control measures. However, some weeds and air-borne cereal pests are relatively unaffected by rotation or other cultivation practices, and the use of agrochemicals has become standard practice, both in rotational and continuous cereal growing systems. Intensive arable farming, with large areas of cereal monoculture, has encouraged the development of cereal crop weeds, pests and diseases. The intensification of cereals has resulted in a build-up of species difficult to control, selected out by continuous agrochemical use, leading to the development of resistance, primarily of diseases, but also, more recently, of weeds. This requires the continued development of agrochemicals, to combat new resistant types. It is likely, unless major advances in plant breeding for resistance to pests and diseases occur, that the production of novel agrochemicals to combat the ability of unwanted species to develop resistance will have to continue for the forseeable future in the present cropping systems. The development of resistance to herbicides by weeds is a much slower process, and there is a wider range of chemicals able to control each of the major cereal weeds than is known to control pests and diseases. Weeds do not spread as fast or as far as pests and diseases, and may be more susceptible to cultivation and rotation changes.

Recently, growth regulators have come into prominence in cereal growing. These are a group of chemicals which have an effect on cereal growth so as to benefit the structure or yield of the plant. One of the major uses is the control of lodging, thus reducing the risks associated with higher nitrogen inputs on cereals, possibly benefiting yield and grain qualities, and reducing harvesting losses. The direct effect on yield, which has been seen experimentally, is more difficult to explain, and is an erratic phenomenon. The first growth regulator used in cereals, chlormequat, was developed in 1959 for the control of lodging, through shortening and strengthening the stem. Little further development occurred until the mid-1970s, when the potential of chlormequat as a growth modifier for yield benefit was recognised and the emphasis of research on cereal growth regulators was changed. It is an area of development still in its infancy, and major developments could occur in the near future.

The Chemicals

Modes of action

The modes of action of agrochemicals can be divided into:

1. Those which prevent the weed, pest or pathogen from reaching or affecting the crop, which can be termed "preventive" materials. They need to be persistent to prevent attack over a period of time, and include: residual herbicides, which control weeds as, or before, they emerge and compete with the crop; fungicides that prevent pathogens from entering host tissue, such as some seed treatments and protective sprays; and pesticides which prevent pests from reaching or damaging the host tissue, such as slug baits.

2. Those which move through the crop plant to control growth, as in the

case of growth regulators, or to control pathogens or pests living on the host plant. These are termed "systemic" or "translocated" chemicals.

3. Those which are translocated within weed plants to sites of action which may be remote from the part of the plant receiving the chemical, such as an underground rhizome system, or may be hidden from the spray, such as a growing point within a cover of leaves.

4. Those which have contact action, killing weeds, pathogens or pests by direct action.

Some chemicals may act in a variety of ways, perhaps dependent on how they are applied to the target. Formulation, spray droplet size, and timing of application are examples of factors which may influence the mode of action of an agrochemical. Many products are formulated as mixtures or recommended for use in tank-mixes, to combine more than one mode of action.

TABLE 4.1 Modes of action of crop protection chemicals

Preventive:

Herbicides – pre-emergence residuals, e.g. pendimethalin, trifluralin
Fungicides – some seed treatments, surface protective sprays, e.g. captafol
Insecticides – soil treatments, e.g. fonofos granules, some protectant seed treatments such as chlorfenvinphos, and foliar protectant sprays which are picked up before the pest reaches the host plant

Translocated/systemic in crop:

Growth regulators – foliage sprays to modify plant growth, e.g. mepiquat chloride
Fungicides – some seed treatments move into the seed, or are translocated within the growing plant, e.g. imazalil, ethirimol, carboxin; others move in the plant from foliage sprays to sites of infection, e.g. benzimidazoles, pyrazaphos
Insectides – foliage-applied chemicals move through the plant to sap-sucking insects, e.g. dimethoate, or through leaf tissue (translaminar), to sucking insects on the opposite leaf surface untouched by the leaf spray, e.g. pirimicarb.

Translocated/systemic in weeds:

Herbicides – foliage sprays with selective action on weeds, moving through plants to site of action, e.g. hormone herbicides (MCPA, mecoprop), wild-oat herbicides (diclofop-methyl, difenzoquat)

Contact action:

Herbicides – foliage sprays with direct action on small weeds; no translocation, e.g. ioxynil, dinoseb
Fungicides – non-penetrant fungicides, killing fungus and diseased tissue on foliage (eradicant), e.g. dithiocarbamates such as maneb, thiram
Insecticides – sprays which directly affect the insect, whether on the ground or the host plant, e.g. synthetic pyrethroids, carbamates, or insects within the ground, e.g. fonofos

Formulation of crop protection chemicals

Crop protection chemicals can rarely be marketed as plain active ingredients. They normally require "formulating" with other materials, to make commercially acceptable products, which will be easy to handle, will not deteriorate during a reasonable storage period, and will not react with materials used for containers or application equipment.

The formulation of an active ingredient may affect its biological activity through altering its selectivity, the mode of action in the organism or its active life on the target.

There are some common features in the formulation of different types of crop protection chemicals. For example, preventive treatments of fungicides and insecticides, and foliage-applied herbicides that only enter the plant slowly, require good covering and "sticking" properties. Materials are added to the formulation to confer these properties. Where an extended residual activity is required from a relatively soluble or volatile soil-applied chemical, formulation as a granule may have the desired effect, whether it is an insecticide, fungicide or herbicide. The use of microencapsulation is being developed as a means of achieving the slow release of a number of different chemicals in or on soil.

The types of formulations developed for crop protection chemicals used in cereals can be grouped into the different types discussed in the following paragraphs (and see Table 4.2, p. 66).

Water soluble solids. Some herbicide active ingredients can be readily formulated as a water-soluble salt. These may be sold simply as solids, such as water-soluble powders (e.g. MCPA), or pellets (e.g. TCA), or in aqueous solution.

Liquids. Most crop protection chemicals for spray treatments are available as liquid products. If they are readily water-soluble, herbicides are frequently sold with water as the main carrier. No fungicides, insecticides or growth regulators are presented in this form.

Non-water-soluble materials, which require to be sprayed, have to be prepared in a way which allows them to be applied in water. If the chemical dissolves in an organic solvent, suitable emulsifying agents can be added so that, when mixed with water, a stable oil-in-water emulsion is formed. A wide range of crop protection chemicals are prepared as emulsifiable concentrates.

An increasing number of non-water-soluble chemicals are becoming available as "flowable" or suspension concentrates. These contain very finely divided active ingredients, mixed with surface-active agents, and dispersed in water give stable suspensions. They have the advantage over wettable powders in not giving rise to problems of dustiness. The size of individual chemical particles can be made smaller than in wettable powders, and this may modify the biological activity of these preparations.

Wettable powders. A wettable powder is prepared if an active ingredient does not dissolve readily in standard solvents. It consists of finely ground chemical, plus a suitable filler, such as clay, also finely ground, together with agents which allow the dispersal and suspension of the fine particles when mixed with water. Some fungicides and herbicides are still sold in this form, but wettable powders are being steadily replaced by suspension concentrates and water dispersible granules, which overcome many storage and spraying problems.

Water dispersible fine granules or grains. These formulations are being developed to allow the suspension of active ingredients which do not mix with water as liquid formulations, (e.g. pirimicarb). Incorporated surface-acting agents provide a spray suspension similar to that produced from a wettable powder.

Granules. Granules contain the active ingredient in a inert base. They are generally designed to improve the residual effect of some of the active ingredients, (e.g. tri-allate, fonofos and chlorfenvinfos), as well as improving handling and application properties. Tri-allate, formulated as a granule for use in cereals in the UK, allows this herbicide to be applied without incorporation, which is necessary if the liquid formulation is used because of the volatile nature of the chemical. The granules are activated by soil moisture.

Bait pellets. A number of pesticides are prepared in the form of bait pellets to attract the target to the active ingredient. For example, the molluscicides methiocarb and metaldehyde are based on wheat or barley meal and bran pellets respectively. Gamma-HCH is available based on bran pellets for the control of leatherjackets. These pellets may be applied overall or in seed drills.

Seed treatments. Fungicide and insecticide treatments may be formulated as dust, powders, liquids and, in the near future perhaps, as microencapsulated preparations. Dust and powders are based on mineral carriers, such as clays and silica, whilst liquid formulations are similar to the liquids used for conventional sprays.

Effect of formulation on biological activity

Various aspects of formulation will affect the biological activity of the active ingredient. In the case of liquid formulations applied in water, the addition of wetting agents, stickers and humectants to the basic formulation of the product can have major effects on the biological activity of a product. Some fungicides and herbicides, (e.g. difenzoquat), are very ineffective unless sufficient wetting agents are added to the formulation. The additives may increase the area of cover of a spray, by breaking down spray droplet surface tension, and so assist uptake of the active ingredient. Wettable powders, water dispersible granules and liquid seed treatments also contain additives to assist adhesion, spread and uptake of the active ingredient into the target.

TABLE 4.2 Formulations available for various application systems

Spray systems:
Water soluble powders and granules
Wettable powders
Suspension concentrates
Emulsion concentrates (oil-in-water or water-in-oil)
Water dispersible granules
Solid application systems:
Granules
Bait pellets
Dusts and powders
Seed treatments:
Dusts and powders
Water soluble powders
Wettable powders
Suspension concentrates
Emulsion concentrates

Differences in target require differences in formulation. For example, penetration of insect exoskeleton may require different chemical and physical properties from those required for the penetration of plant tissues. In some cases, a persistent protective presence of the insecticide or fungicide on the surface of the crop plant is required, and additives are used in the formulation to extend the active period of the spray for as long as possible. Some fungicides are required to penetrate fungal tissues on the surface of the crop plant and may need different formulating agents than those fungicides which are required to penetrate the plant tissues.

A number of herbicides are selective in action between crop and weeds because of differential retention caused by differences in leaf shape and size, foliage orientation and the properties of the leaf surface. The type of selectivity can be changed markedly by formulation differences, as well as by the drop size characteristics of the spray.

Surfactant materials, including wetting agents, also affect the pattern of spray droplet formation, and consequently the evenness of deposition. This is particularly important for contact-acting crop protection chemicals, which are restricted in action to the zone around the point of deposition. The optimum deposition pattern for contact products will vary with the size of target. A translocated material moves throughout much of the target, so droplet deposition patterns are sometimes less critical, although the deposition requirements for achieving adequate uptake into the target may be just as critical as those needed for effective contact activity.

In developing a formulation, the manufacturer of a crop protection product will endeavour, as far as possible, to maximise the useful activity of the product and minimise any inconvenience to the user. Some label recommendations require the addition of extra wetters or other adjuvants for certain uses. Such recommendations should be closely adhered to, since the use of the wrong type or amount of additive may affect the formulation, as also can mixing with other crop protection chemicals; rain-fastness, selectivity and possibly activity may be drastically changed and the toxicological characteristics of a product may be affected.

The herbicide manufacturer will not necessarily have covered all possible combinations of additives. Independent research may well discover some further beneficial additions to a formulation. However, such additions must be carefully monitored, in case some adverse changes to the other aspects of activity have occurred, and they should not be used until proved satisfactory from all points of view, including the requirements of Schemes and Acts protecting the safety of the user and the environment.

Crop safety

The safety of agrochemicals to crops varies considerably. The sensitivity of a crop protection chemical is usually only a matter of degree, and overdosing will kill the crop as well as the pest, pathogen or weed. For each chemical, there will be rates and crop growth stages at which use is safe. Frequently there is only a fine balance between the dose of chemical used and the stage of the development of the crop, pest, pathogen or weed.

As herbicides are designed to control plant species that are often very similar to the crop they infest, it is to be expected that damage from herbicide applications would be more frequent than that from fungicides or insecticides, and this is borne out in practice. Many herbicides are only recommended for use in some, not all, cereal species, (see Tables 4.4, 4.5 and 4.6); some are only tolerated by certain varieties within a cereal species.

Spray timing and the condition of the crop are critical for the application of most foliage-applied herbicides. A weak crop may succumb to the effects of normally safe treatments, or be rendered more sensitive to the effects of certain diseases, or to the effects of frost following treatment. Under certain conditions, usually those of high rainfall and in very light soils, soil-applied herbicides may also cause considerable crop damage.

Fungicides are generally safer to cereals than are herbicides, but seed treatments can occasionally retard germination, and crop scorch and check may also occur from foliage applications. However, foliage damage more usually occurs when non-compatible mixtures with herbicides are used, and the combined effect aggravates the potential damage by the herbicides. Mixtures with liquid fertiliser are particularly prone to cause severe leaf scorch.

Insecticides also tend to be safer to the crop than herbicides, but insecticidal seed treatments may also affect germination, and some insecticides may be quite phytotoxic.

Growth regulators have been known to cause damage, and reduce yields, largely due to mistiming of applications. As they interfere with the growth of crop plants, it is to be expected that, under certain circumstances, such interference may not be beneficial.

User and environmental safety

At the time of writing, new legislation is being enacted in the UK to replace the current non-statutory Pesticide Safety Precautions Scheme, which has controlled the introduction and use of pesticide products for over 27 years, and provided an excellent safety record in their use. Enactment of Part III of the Food and Environment Protection Bill will enable orders and regulations to be made to control the import, sale, supply or use of pesticides in the UK. The powers sought in the Bill cover the full range of conditions which might affect safety and efficacy, from label design to application method. The intention is to limit pesticide use to a minimum consistent with reasonable levels of productivity, and to ensure that no pesticide is used unless its use has been demonstrated to be effective, safe to humans and to animals and to present the least possible risk to the environment.

The Health and Safety at Work etc. Act, 1974, and the Poisonous Substances in Agriculture Regulations 1984 place general obligations on employers, employees, self-employed, manufacturers and suppliers to ensure that substances and machines are safe for their intended use. Copies of the regulations are available from Her Majesty's Stationery Office (HMSO), who can also supply a copy of the booklet HS(R) 20 *A guide to the Poisonous Substances in Agriculture Regulations 1984* which explains how the requirements of the regulations can be met. Most of the chemicals used in cereals which are scheduled in the regulations are insecticides, an exception being the herbicide dinoseb. The Poisons Act, 1972, regulates the sale of certain highly toxic chemicals, the one most likely to be used in a cereal growing system being paraquat. The provisions of the Act require that chemicals in the Poisons List Order may only be sold by a retail chemist or "listed seller of poisons" on the production of a signed order or to a person signing the poisons book in the shop.

A particular hazard arising from the spraying of chemicals is the drift of herbicide. This is frequently implicated in damage to the flora of hedgerows, gardens, neighbouring crops and other areas of environmental interest. Some herbicides may affect pollinating insects, but, in general, together with growth regulators and fungicides, are considerably less toxic to animal life than insecticides. Nevertheless, maximum care in the handling and use of all agrochemicals is recommended and the instructions for safe use, which are incorporated on product labels or in the technical literature which should accompany the sale of all crop protection chemicals, should be followed implicitly.

Herbicides

Chemical groups and properties

Over 200 herbicide products are available for use in cereals in the UK. Most are designed for use in wheat and barley; rather fewer for oats and undersown cereals and a smaller number for use in rye, triticale and durum wheat, (see Tables 6.2, 6.3 and 6.4). This simply reflects the relative importance of these crops in the world and UK markets. The British Crop Protection Council *Weed Control Handbook: Principles*, 1982 7th Edn, Blackwell, Oxford, examines the herbicides and their biological properties in some detail.

The section below examines the types of herbicides available according to their type of activity and gives examples of the types of weeds controlled. More detailed information can be found in product literature, Ministry of Agriculture Fisheries and Food Booklets 2252 and 2253 *Weed Control in Cereals Spring and Summer*, and *Weed Control in Cereals Autumn and Winter* which are revised annually and are also available as spring and autumn supplements in the *Farmers Weekly* (Business Press International, Sutton), and in The British Crop Protection Council *Weed Control Handbook Vol. II Recommendations*, 1978 8th Edn, Eds. J.D. Fryer and R.J. Makepeace, Blackwell, Oxford.

Herbicide types

The range of herbicides can be divided broadly into those which are only taken up by foliage, and those which have a residual effect, that is they persist in the soil for a sufficient length of time either to be taken up by the roots of established plants, or to control weeds which germinate some time after the chemical has been applied.

Some residual herbicides may be taken up also by plant foliage, but to facilitate grouping, all those herbicides where uptake is largely *via* the soil are included as residual herbicides. There are 16 herbicides with residual activity used in cereal crops in the UK at the present time.

The selectivity of residual herbicides may be dependent on the relative depth of the crop and weed seeds in the soil, on inherent selectivity in the crop, or on differences in the uptake by the crop and weed roots or growing shoots.

Pre-emergence residual treatments are applied before the weed emerges, and usually before the crop emerges, killing weeds as they germinate or grow through the chemical layer in the soil. These chemicals usually have a relatively low water solubility, so persist in the upper layer of the soil where the majority of annual weeds germinate. The crop sown beneath this layer may be unaffected during germination.

Residual herbicides which are applied post-emergence of crops and weeds may be inherently tolerated by the cereal plant, or be taken up selectively by the weed roots and foliage.

Some residual herbicides can be applied either pre- or post-emergence of crops and weeds (e.g. isoproturon, chlorsulfuron), whereas others may only be applied pre-emergence, (e.g. terbutryne), or post-emergence, (e.g. metoxuron).

Herbicides which are taken up by the foliage may be sub-grouped into the contact herbicides, which affect only that part of the plant in direct contact, or very close to the point of contact, with the herbicide and the translocated herbicides which, once taken in by the plant, may be transported or moved to the sites of herbicidal action.

The selectivity of contact herbicides is dependent largely on differential retention and uptake by the foliage of the crop and weed, although there may also be some inherent selectivity involved.

Certain translocated herbicides are termed growth regulators, or, in common parlance, "hormone" herbicides, because of their disruptive effects on the hormone-controlled growth regulating systems within the plant. Use of the term "hormone" avoids confusion with the growth regulator crop protection chemicals discussed later. The original hormone herbicides were developed in Britain and the United States of America through a search for chemicals which would regulate the growth of plants. Currently, 11 hormone herbicides are used in UK cereal crops. The discovery of hormone herbicides is described by C. Kirby in *The hormone weedkillers*, 1980, published by The British Crop Protection Council, Croydon.

At certain cereal growth stages, depending on the herbicide involved, the crop is vulnerable to otherwise safe hormone herbicides. The periods during which the leaf, ear and grain primordia are formed, and the period of anthesis, are particularly vulnerable stages. (See Appendix I).

Other translocated foliage-acting herbicides used in cereals are not of the hormonal type, although some do act in a similar manner. In particular, the broad-leaved herbicide, benazolin, appears to act as an auxin-type growth regulator and the grass weed herbicide, diclofop-methyl, has anti-auxin type activity. Others have an effect on other biochemical and physiological pathways in grass weeds.

Product types available

The products available for use in cereals may be made up of one or more herbicides which individually have residual, or contact, or translocated activity. Such products combine these individual effects to give a broader range of activity.

The mixing of herbicides is becoming very complex as agrochemical manufacturers attempt to provide the farmer with products that will combat very broad spectra of weed populations at different growth stages, and a number of mixture products containing 3 or 4 herbicides are now available. Some of the novel combinations include residual with foliar-acting herbicides, and this, combined with the fact that certain residual herbicides may be used both pre- or post-emergence, is blurring the once clear distinction between different herbicide types. For this reason, the use of the following terms to distinguish cereal herbicide product types is suggested:

Residual;
Contact foliage-applied;
Translocated foliage-applied;
Residual plus contact;
Residual plus translocated;
Residual plus contact plus translocated;
Contact plus translocated.

Products combining different herbicides have been developed to save the farmer the trouble of mixing the herbicides himself. However, there has been an increasing trend to on-farm mixing of individual herbicides and, in response, the pricing of manufactured mixtures by the agrochemical companies is becoming more competitive.

To develop such mixtures, one has to know the weeds controlled by individual components, their persistency, and crop safety.

Residual herbicides

Residual herbicides may be applied:

pre-emergence only of crop and weed, (e.g. terbutryne);
pre- or post-emergence of crop but pre-emergence of weeds, (e.g. trifluralin);
pre- or post-emergence of crops and weeds, (e.g. isoproturon);
post-emergence of crop, pre- or post-emergence of weeds, (e.g. metoxuron).

This picture is complicated further because the timing of the application may be crop and/or variety dependent, and weed dependent. Current individual product literature must be referred to for full details.

Most residual herbicides are used in autumn cereals. Some of them have sufficient persistence to affect spring germinating weeds after an autumn application.

In a well-established winter barley crop, ground cover is usually such that, in the spring, few weeds will emerge to cause crop losses, but in poorly established barley crops and the more open wheat, rye and triticale crops, spring germinating weeds may be as much of a problem as autumn germinating weeds. Residual herbicides used in the autumn in these crops require a long period of persistence to control spring germinating weeds satisfactorily.

The relative persistence in mineral soils at phytotoxic levels of the residual herbicides used in cereals varies as shown in Table 4.3. The safe period between application and sowing succeeding crops varies considerably according to crop, soil and climatic conditions.

TABLE 4.3 Persistence in soil of residual herbicides used in cereals

Less than three months:	*Three–six months:*	*Six months or more:*
bifenox	chlortoluron	chlorsulfuron
cyanazine	isoproturon	metsulfuron-methyl
metoxuron	linuron	pendimethalin
terbutryne	monolinuron	trifluralin
	methabenzthiazuron	isoxaben
	tri-allate	
	trietazine	

Pre-emergence of crops and weeds

Linuron and mixtures. Linuron may be used alone where common chickweed, *(Stellaria media)*, and mayweeds, *(Matricaria* and *Chamomilla* spp.), are expected to be the main problem in winter crops, but for broader spectrum weed control, product mixtures with bifenox or monolinuron or trifluralin (plus trietazine) are used.

They are recommended on autumn-sown wheat and barley. The products are designed for a wide range of broad-leaved weed control, with some effect on meadow-grasses, *Poa* spp.). There are differences in their weed spectra, the most important being the mixture with bifenox to control field pansy *(Viola arvensis)* and cleavers *(Galium aparine)*. Bifenox inhibits seed germination and early seedling growth; persistence of effect is about 2–3 months and some initial crop scorch to barley is often seen. The urea herbicides, linuron and monolinuron, act largely by inhibiting photosynthesis, so that emerging weeds are pale or yellowing, then quickly necrose. They can persist in the soil for 3–6 months, but under wet and mild autumn conditions their active life may be reduced. The addition of trifluralin extends the persistency of a product based on linuron to 6 months or more under cool conditions, but its weed spectrum is somewhat complementary to that of linuron, and certain weeds may grow away once the effect of the linuron has worn off.

Trifluralin is absorbed by the shoot and root and inhibits cell division, so susceptible weeds may either fail to emerge, or are very stunted; alone it only has pre-emergence activity, but mixtures with foliage-uptake herbicides have been developed.

Pendimethalin. This is generally used alone and has a very broad spectrum of activity on broad-leaved weeds, plus activity on black-grass

(Alopecurus myosuroides) and meadow-grasses. Pendimethalin has a similar mode of action to trifluralin, inhibiting cell division, thus preventing seedling development. It is very persistent, and most spring germinating broad-leaved weeds are controlled from an autumn treatment; late germinating mayweeds are an important exception. It can be used on winter wheat, winter and spring barley and on rye, pre- or post-emergence of the crop but foliage-acting herbicides have to be used with it if weeds have emerged.

Terbutryne. This triazine acts mainly through root uptake by blocking respiratory and photosynthetic mechanisms. Affected weeds, if they emerge, quickly die back. It can be used on all winter-sown cereals and has a fairly broad spectrum of broad-leaved weed control, especially common chickweed, mayweeds and meadow-grasses, plus some activity on black-grass and perennial rye-grass *(Lolium perenne)*. Cleavers and speedwells are not well controlled.

TABLE 4.4 Residual herbicide use in cereals

	Winter						Spring		
Crop	Wheat	Barley	Oat	Rye	Triticale	Durum	Wheat	Barley	Oat
Herbicide:									
bifenox	√	√[1]	√[1]	×	×	×	×	√[1]	√[1]
chlorsulfuron	√[2]	×	×	×	×	×	×	√[4]	×
chlortoluron[5]	√	√	×	×	√	√	×	×	×
cyanazine	√	√	√	×	×	×	√	√	√
metsulfuron-methyl[3]	√	√	√	×	×	×	×	×	×
isoproturon	√	√	×[6]	√	√	√	×[5]	×[6]	×
isoxaben	√	√	√	×	√	√	×	×	×
linuron	√	√	×	×	×	×	×[7]	×[7]	×[7]
methabenzthiazuron	√	√	√	√	√	√	×	√	×
metoxuron + simazine	√	√	×	×	×	√	×	×	×
pendimethalin	√	√	×	√	×	×	×	√	×
terbutryne	√	√	√	√	√	√	×	×	×
tri-allate	√	√	×	×	×	?[8]	√	√	×
trifluralin[9]	√	√	×	×	×	×	×	×	×

√, label recommendation available; ×, no recommendation.

Notes:

1. Product mixture with chlortoluron limited to certain varieties of winter wheat only. Product mixture with linuron only recommended on winter wheat. Recommendation is for mixture with mecoprop.
2. Product mixed with methabenzthiazuron.
3. Product mixed with chlorsulfuron.
4. Product mixed with ioxynil-bromoxynil.
5. Certain varieties only.
6. Mixtures of other herbicides with low amounts of isoproturon are recommended.
7. Mixture with 2,4-DB/MCPA is recommended for use in undersown spring cereals. Linuron is available alone and in mixture with bifenox or mondinuron or trifluralin.
8. Trials indicate crop tolerance, but no label recommendations made.
9. In mixture with linuron.

The table indicates crop tolerance at commercially recommended rates; in some cases tolerance depends on the lowest of several rates being used.

Pre- and/or post-emergence of crop and weeds

Bifenox apart from being used pre-emergence in winter wheat and barley in mixture with linuron, is formulated for use post-emergence in wheat and barley in mixture with the translocated foliar herbicide mecroprop, and with chlortoluron for pre- or post-emergence use in certain wheat varieties. It is known to scorch barley, but there is little evidence of effect on yields. It appears to be taken up by leaves as well as roots and in addition to inhibiting seedling growth, it has some direct contact action. It has a fairly broad spectrum of weed control with a particularly useful advantage in controlling common chickweed, field pansy and cleavers. The addition of mecoprop extends the broad-leaved weed control spectrum and a greater than additive effect in activity from the mixtures is claimed; but mayweeds, *Polygonum* spp. and grasses are not well controlled. The mixture with chlortoluron extends the activity on grass weeds.

Chlorsulfuron is used in pre- or post-emergence in winter wheat in mixture with the residual herbicide methabenzthiazuron and post-emergence in late-sown winter wheat and barley, and in spring wheat and barley, as a mixture with ioxynil and bromoxynil contact herbicides. A mixture with metsulfuron-methyl can be applied pre- or post-emergence in winter oats and winter wheat and post-emergence in winter barley. It prevents cell division and elongation, and weeds appear stunted or do not emerge. It is particularly effective on seedling weeds and controls a very broad spectrum of broad-leaved weeds plus some grasses at high rates. The mixture with metsulfuron-methyl gives a general extension of broad-leaved weed control. Methabenzthiazuron has been added primarily to improve grass weed, field pansy and red dead-nettle *(Lamium purpureum)* control in the autumn, and ixoynil/bromoxynil to improve the control of *Polygonum* spp. in the spring. Applied at the autumn rate, persistence in the soil for up to one year may be expected. This may have consequences on following crops in the rotation, and recommendations on following crops in product literature should be carefully studied before use.

Metsulfuron-methyl is a herbicide with similar properties to chlorsulfuron for post-emergence use in winter wheat, and certain varities of winter barley and spring barley. A wide range of broad-leaved weeds are controlled, but mixtures are required for the control of cleavers, some speedwells *(Veronica* spp.*)*, and to improve *Polygonum* spp. control. A mixture with chlorsulfuron improves the control of speedwells, mayweeds and field pansy in particular.

Chlortoluron is a urea herbicide mainly taken up by the roots, although there may be some foliar uptake. Inhibition of photosynthesis is considered the primary mechanism of action, although not necessarily the only mechanism. Affected plants pale, or go yellow before dying. A broad spectrum of seedling broad-leaved weeds are controlled in winter wheat and barley and certain varieties of triticale and durum wheat pre-emergence only, but it is particularly useful for the control of seedlings of most annual grass weeds and the seedlings of some perennial grass weeds. Certain important autumn germinating broad-leaved weeds, notably cleavers, red dead-nettle, field pansy and speedwells are not controlled, and mixtures or sequences with other products will be required when these weeds are present. There are also considerable cereal varietal restrictions in the use of chlortoluron.

Cyanazine is generally used post-emergence in spring on winter and spring-sown wheat, barley and oats, in mixture with the translocated foliar herbicides mecoprop or MCPA; or in mixture with bentazone/dichlorprop in wheat and barley. These mixtures give a very broad spectrum of broad-leaved weed control based on the extensive weed control of bentazone/dichlorprop, mecoprop or MCPA, plus the good activity of cyanazine on, in particular, chickweed, red dead-nettle, mayweeds and speedwell. The safe periods for using the mixtures depend mainly on the herbicides mixed with the cyanazine and vary with cereal type and species. Cyanazine may also be used alone, pre- or post-emergence, in the autumn in wheat or barley; in particular to control chickweed and seedling annual meadow-grass. Cyanazine is a triazine and as such, its main site of uptake is *via* the roots, but it is relatively soluble in water, and some foliar uptake probably takes place. The respiratory and photosynthetic systems of susceptible plants are inhibited, leading to chlorosis of leaves with general inhibition of growth, followed by necrosis. Cyanazine has a relatively short life in the soil and, under very moist conditions, may not persist for more than a few weeks.

Isoproturon is used alone, pre- or early post-emergence, in all varieties of winter wheat and barley, and pre-emergence in rye and certain varieties of triticale. It is frequently used as an ingredient of post-emergence mixtures for very broad spectrum weed control. Such mixtures include the addition of ioxynil and bromoxynil, and mecoprop plus ioxynil. There is also a mixture with trifluralin for pre-emergence and early post-emergence use. Isoproturon alone has useful activity on a limited range of seedling broad-leaved weeds including chickweed, mayweeds and corn marigold *(Chrysanthemum segetum)*, plus good activity on black-grass, meadow-grasses, other annual grasses and seedlings of perennial grass weeds. Its mode of action is similar to that of chlortoluron, with the roots being the primary site of uptake, but some foliar uptake may take place. Symptoms appear as chlorosis, followed by necrosis, of the leaves. It is more soluble than chlortoluron and tends to have a shorter life in the soil, but, under suitable conditions, its residual activity may last for about 3 months.

Methabenzthiazuron is used alone pre-emergence in winter wheat, barley, oat, rye, triticale, durum wheat, and spring barley, also post-emergence in winter wheat and named varieties durum wheat, up to 6 weeks after drilling, provided the weeds are at the seedling stage. Mixtures with chlorsulfuron and terbutryne are also available Methabenzthiazuron controls a limited range of broad-leaved weeds, and moderate levels of black-grass and meadow-grasses, and the seedlings of some other annual grasses. The mode of action is primarily *via* the soil with root uptake leading to yellowing and stunting of seedlings followed by necrosis, or a failure of the weeds to emerge. Phytotoxic levels of methabenzthiazuron persist in most mineral soils for 3–4 months.

Metoxuron is only used post-emergence in some varieties of winter wheat, durum wheat and barley. It is primarily used where there is a mixture of black-grass with broad-leaved weeds, but there is some effect on wild-oat *(Avena fatua)* seedlings, and the seedlings of some other annual grasses. It is also used in sequence with tri-allate for barren brome *(Bromus sterilis)* control. A small quantity of simazine is added to one metoxuron-based product, but this has little effect on the spectrum of weed control. Metoxuron, in common with urea herbicides, is taken up primarily by the roots, but some foliar activity also takes place. Paleness and chlorosis of susceptible plants occurs, followed by necrosis. Some contact scorch can also occur.

Tri-allate controls wild-oat, black-grass and other annual grasses, pre-emergence, in all varieties of wheat and barley. A granular formulation of the herbicide may be applied either pre- or post-emergence, but only small wild-oats are controlled post-emergence. A range of sequences with other grass herbicides are recommended to improve the control of grass weeds. Due to its volatility, the liquid formulation has to be incorporated as soon as possible after application. Absorption appears to be greater *via* the emerging seedling shoots than the roots in most susceptible species. There is general inhibition of growth and the grasses often fail to emerge. Phytotoxic levels of tri-allate may persist in the soil for 3–6 months. It is not active in dry conditions but may be re-activated when soil moisture levels are sufficiently high. A mixture of tri-allate and chlortoluron in granular form which improves activity on blackgrass and extends weed control to include some broad-leaved species is available for use in winter and spring wheat and barley.

Isoxaben controls a wide range of broad-leaved weeds in all winter cereals, pre-emergence of the weeds. Uptake from the soil is *via* the roots and low solubility extends its residual activity in soil to at least 6 months. The chemical inhibits cell division in susceptible plants.

Soil conditions and residual herbicide activity

Residual herbicides require reasonable levels of soil moisture for activation. Early autumn and spring applications occur in relatively dry periods of the year, and may result in a considerable reduction in activity. Activation as soil conditions improve may not be sufficient to control weeds which have already emerged, although there may be some residual effect on emerged weeds, if they are already in a stressed condition.

Most residual herbicides do not work well in soil with an organic matter content greater than 5-10%, (by chromic acid test), or on heavy clay soils. They are, to a varying degree, absorbed onto organic or clay particles and are thus rendered less available to the plants. As a result, the recommended doses for these herbicides may vary according to soil type, or it may be recommended that they are not used on highly organic or clay soils.

There is increasing evidence of reduced activity of residual herbicides in minimum cultivation or direct-drill situations, particularly when they have taken place in conjunction with stubble burning over a number of years, because of the build up of ash and organic residues at the soil surface. The urea herbicides have been particularly affected by such surface conditions, and ploughing is suggested when the build up of organic residues is substantial.

Very sandy soils can also affect the activity and safety of residual herbicides. Under very wet conditions, in soils where there is a very small clay or organic fraction, movement of otherwise fairly insoluble chemicals down the soil profile may occur. Furthermore, the clay and organic particles present near the surface may also be washed down the soil profile, taking absorbed chemicals down with them. This may reduce the control of shallow germinating weeds.

Very sandy or gravelly soils may also reduce the crop safety of residual herbicides which depend on their low mobility in the soil, and their consequent separation from the crop seed. Movement of the herbicide down the profile in very wet conditions can take the herbicide down to the crop seed and root zone, with resultant damage to the crop which normally would not come into contact with the herbicide at the susceptible stage. None of the herbicides applied pre-emergence of the crop appear completely immune from causing damage under such conditions, but the mechanism of their movement through the soil profile is inadequately understood. There does not appear to be any crop cultivar factor involved. All cereal seed should, however be sown at 25-35 mm depth if a residual herbicide is to be used, to minimise the chances of damage occurring, whatever the soil type.

In all cases, these materials require good seedbed preparation, with a fine firm tilth; fluffy seedbeds should be consolidated. The presence of clods and stones can reduce weed control, because they shield areas of soil and germinating seedlings from the spray.

Certain residual herbicides, notably chlortoluron and metoxuron, are not safe to all varieties of wheat and barley, because either those varieties do not have the ability to detoxify the herbicide within the plant, or they retain the chemical on their foliage, or both factors operate.

Some of the herbicides are tolerated by certain cereals but not others. For example herbicides which have activity on wild-oat are unlikely to be safe on oat crops. Those with activity on rye-grasses are less likely to be safe to rye and triticale. Other herbicides are not recommended on certain cereals because insufficient testing has been carried out.

Contact foliage-applied herbicides

TABLE 4.5 Contact herbicide use in cereals

	Winter						Spring		
Crop	Wheat	Barley	Oat	Rye	Triticale	Durum	Wheat	Barley	Oat
Herbicide:									
bromoxynil/ioxynil[1]	√	√	√	√	√	√	√	√	√
dinoseb acetate	√	√	√	√	×	×	√	√	√
dinoseb amine or ammonium salt	√	√	√	×	×	×	√	√	√

1. Some products do not include oats or only recommended use on oats in spring.

Bromoxynil and **ioxynil** are hydroxybenzonitriles, which interfere with vital processes within the cells close to the point of uptake. Translocated activity is very limited; consequently, good spray cover of the foliage is essential and the best control of weeds is achieved when they are small. A rapid yellowing or chlorosis and scorch of the foliage of susceptible plants occurs. There is no useful residual activity in the soil and only broad-leaved species are controlled, although some transient scorch of grasses may occur. Temperature and light affect the activity of the hydroxybenzonitriles. Activity is most rapid and effective under bright light and warm conditions.

The combination of ioxynil and bromoxynil controls most of the broad-leaved weeds found in cereals at the seedling stage. Once the weeds become larger than 4-8 leaves, depending on the species, then activity becomes variable, and translocated herbicides are added to improve activity. For example, for broad-leaved weed control, mixtures developed include:

ioxynil + mecoprop;
ioxynil + bromoxynil + mecoprop;
bromoxynil + ioxynil + benazolin;
ioxynil + benazolin + mecoprop;
clopyralid + bromoxynil + ioxynil + mecoprop.

For controlling grass weeds in addition, mixtures including residual herbicides have been developed, e.g.:

isoproturon + ioxynil + bromoxynil;
isoproturon + ioxynil + mecoprop.

Ioxynil and bromoxynil used alone are safe to cereals over a wide range of growth stages. The tolerance of crops to mixtures at the time of use depends upon the other herbicides added. The lack of residual activity makes ioxynil and bromoxynil particularly suitable for use in cereals prior to undersowing.

Dinoseb, as the acetate, amine, or ammonium salt is a further contact herbicide used in cereal crops. It is a phenol herbicide, producing rapid browning and desiccation of susceptible plants. Dinoseb has been reported to affect many processes within plant cells, but translocation is very limited and good covereage is essential. It can remain active in the soil for a short time.

It controls many broad-leaved weeds, but because of the high mammalian toxicity and the frequency with which crop scorch occurs under warm or frosty conditions, or if heavy rain follows application, its use has diminished. Its main use is in crops undersown with grass and clover or lucerne. There are no product mixtures including dinoseb, and tank-mixes are not recommended. Dinoseb can be used on all cereal crops except triticale and rye, over a wide range of growth stages, starting at the three leaf stage. To achieve good cover, high application volumes are recommended.

Pre-sowing use of contact herbicides

Contact herbicides may be used to clean-up seed beds by controlling emerged weeds before preparations for sowing. Paraquat and paraquat/diquat formulations are most commonly used. They will kill most annual weeds and volunteer cereals and check the growth of some perennial weeds, normally with no residual activity in the soil to interfere with the germination of the crop seed. The translocated herbicide glyphosate can be used in the same way.

Translocated foliage-applied herbicides

Growth regulator or hormone herbicides

The herbicides available in the UK which exhibit growth regulating or hormonal type activity are listed below. Their activity is primarily on annual and broad-leaved weed species.

2-(Aryloxy) alkanoic acids

MCPA
2,4-D
mecoprop
dichlorprop.

Precursors of phenoxyacetic acids

MCPB
2,4-DB.

Arylcarboxylic acids and their esters

2,3,6-TBA
dicamba
clopyralid.

Other miscellaneous herbicides

benazolin
fluroxypyr

These herbicides are mainly absorbed through the foliage, but some root uptake may occur. The activity is primarily on broad-leaved weed species, and their effects include epinasty of leaves, stems and petioles, stem distortion and splitting, deformities of flowering structures, and the development of root primordia above ground. Necrosis develops slowly. Their activity is improved under good growing conditions.

Cereals may be affected by these herbicides if treated outside certain growth stages. At early growth stages, before ear initiation, when plants have 1 or 2 leaves, leaf and tiller deformities may result. If some products are applied at ear initiation stages, ear distortions such as missing florets and "forked" ears are likely to occur. Applications after booting may lead to sterility of the ears giving "rat-tailing" symptoms, due to the failure of grain to develop. MCPA, 2,4-D, and 2,3,6-TBA are the most usual materials to be associated with such symptoms, and are recommended to be applied only from the leaf sheath erect stage of winter cereals (GS30), or the 5 leaf stage of spring cereals (GS15), up to the first node detectable stage (GS31). Under good growing conditions, the cereal plant may pass through these growth stages in as little as a week, which limits the usefulness of these herbicides.

Mecoprop, dichlorprop, clopyralid and fluroxypyr are more flexible in their use, and products containing these hormone herbicides are applied over a wide range of growing stages. Mecoprop is now recommended for use up to the 2 nodes detectable stage (GS32) and fluroxypyr up to flag-leaf emergence (GS39) in winter cereals.

Some product mixtures have broader timing recommendations than some of the individual components. This is largely due to the lower amounts of the more harmful components being used in the mixtures than would be used alone.

The herbicides vary considerably in the spectrum of broad-leaved species they control at rates suitable for use in cereals, although many of the herbicides affect most species to a greater or lesser extent. The literature previously cited (page 69) lists many of the annual broad-leaved weeds which are controlled by each herbicide and mixtures of them.

In general, **MCPA** is particularly useful for the control of *Brassicae* weeds, common poppy *(papaver rhoeas)*, and hempnettles *Galeopsis* spp., and suppresses perennial broad-leaved weeds such as creeping thistle *(Cirsium arvense)*. **2,4-D** also has activity on perennial broad-leaved weeds, and may provide temporary suppression of speedwells *(Veronica spp.)*.

Mecoprop has good activity in common chickweed and cleavers, and has formed the basis of many winter cereal herbicides. There have been some reports of poorer than expected activity on cleavers from the east of England and on heavier soil types. The reason for this is unclear, but it may be due to the erratic germination pattern of this species, or to the increased use in early spring when conditions for good activity are often poor.

Dichlorprop also has useful activity on chickweed and cleavers, and this, with useful activity on black bindweed *(Fallopia convolvulus)*, and redshank *(Polygonum persicaria)*, makes it a useful ingredient of spring herbicide products.

2,3,6-TBA and *dicamba* are only recommended in spring-applied mixtures, and have activity on common chickweed, cleavers, knotgrass *(Polygonum aviculare)*, and redshank. 2,3,6-TBA also has activity on mayweeds.

MCPB and **2,4-DB** are tolerated by legumes, so their main use is in cereals undersown with clover or lucerne. They can be used alone, but have relatively limited activity, and are more usually mixed with small amounts of less clover-safe herbicides such as MCPA or 2,4-D, or with benazolin.

Benazolin is not related to the other hormone herbicides, but it acts in much the same way, producing similar symptoms to 2,4-D. It is tolerated by legumes so can be used in undersown crops, and it is particularly useful for the control of common chickweed and cleavers.

Clopyralid is a recently developed hormone herbicide. It has activity on annual and perennial *Compositae* species, and is made available for use in cereals both alone and in a wide range of mixtures. Crop safety is good, and mixtures with ioxynil, bromoxynil and low rates of mecoprop are recommended for use from the one leaf stage up to, but not including, 3 nodes detectable stage.

Fluroxypyr has particularly useful activity on cleavers of all sizes, chickweed and common hemp-nettle, amongst other broad-leaved

TABLE 4.6 Hormone type herbicides use in cereals

	Winter						Spring		
Crop	Wheat	Barley	Oat	Rye	Triticale	Durum	Wheat	Barley	Oat
Herbicide:									
benazolin based products	√	√	√*	×	×	×	√	√	√
clopyralid[1]	√	√	√	×	×	×	√	√	√
2,4-D*[2]	√	√	√	√	×	×	√	√	×
2,4-DB*	√	√	√	×	×	×	√	√	√
dicamba*	√	√	√	×	×	×	√	√	√
dichlorprop*[3]	√	√	√	×	×	×	√	√	√
MCPA*	√	√	√	√	×	√	√	√	√
MCPB*	√	√	√	√	×	×	√	√	√
mecoprop[4]	√	√	√	×	×	√	√	√	√
2,3,6-TBA*[5]	√	√	√	√	×	×	√	√	√
bentazone[6]	√	√	√	√	√	×	√	√	√
fluroxypyr[7]	√	√	√	√	√	×	×	√	√

√, label recommendation made; ×, no recommendation made; *, use in spring only.

Notes:

1. When used alone. Mixtures with bromoxynil/ioxynil/mecocrop are not recommended for spring cereals.
2. Mixtures cannot be used on rye.
3. Mixtures with 2,4-D cannot be used in spring oat. Mixtures with bentazone can be used in triticale.
4. Many mixtures have different recommendations, e.g. with bifenox cannot be used on spring wheat, and 2,4,6-TBA with dicamba/MCPA can be used in rye.
5. Mixture with dichlorprop and/or mecoprop cannot be used in rye.
6. Available for cereals in mixture with dichlorprop. Usually recommended as tank-mix with cyanazine which can only be used in winter wheat and barley.
7. When used alone. Mixtures have further restrictions.

In some cases tolerance depends on the lowest of several doses being used. Not all recommended combinations are included here; mixtures may be restricted to a different range of cereal types.

weeds. Considerable safety to cereals is claimed, with recommendations for use up to the time of full flag leaf emergence (GS39).

Other translocated foliage-applied herbicides

Bentazone has symptoms of activity similar to the triazines and ureas on many broad-leaved weed species. Photosynthetic inhibition is believed to be the primary mechanism of action, leading to chlorosis and growth inhibition. It has been demonstrated to be readily translocated in susceptible species. Clover and lucerne are not susceptible, so it is used in undersown cereals, usually in a recommended tank-mix with MCPB. A bentazone mixture with dichlorprop is produced for cereals not undersown with clover or lucerne from leaf-sheath erect (GS30) to the 2-node stage (GS32) in winter wheat, barley, oats and rye and from early tillering up to GS32 in spring wheat, barley and oats. It is also recommended to be used in tank-mix with cyanazine, in which case it is used from the 3 leaf stage (GS13) to the leaf sheath erect stage (GS30) on winter wheat and winter barley only. Cyanazine adds residual activity, which is particularly useful against common chickweed. Bentazone is particularly effective on mayweeds, corn marigold, cleavers and chickweed. Dichlorprop improves the activity of the mixture on *Polygonum* spp.

Benzoylprop-ethyl, **flamprop-M-isopropyl**, and **flamprop-methyl**, are closely-related herbicides developed for the control of wild-oats. They are mainly absorbed by the leaves. Cell division and elongation are inhibited in susceptible species, so they appear stunted and darker green in colour.

Benzoylprop-ethyl can be used in all commonly grown varieties of wheat and is safe on crops undersown with grass and clover. Best results are obtained when the wild-oats are well tillered and growing actively. The crop can be treated from leaf-sheath erect (GS30) up to just before the fourth node detectable stage (GS34).

Flamprop-M-isopropyl can be used in all commonly grown varieties of wheat, (including durum), barley and triticale. The crop can be undersown with rye-grass or clover. Wheat can be treated from leaf sheath erection up to the fourth node detectable stage, and barley and triticale up to just before the 3 nodes detectable stage. The wild-oats should be growing vigorously. There is some useful suppression of black-grass, also the aerial growth of false oat-grass or onion couch *(Arrhenatherum elatius* var. *bulbosum)*, and black bent *(Agrostis gigantea)*.

Flamprop-methyl also gives useful suppression of black-grass and other grasses, in all commonly grown varieties of winter wheat, including crops undersown with rye-grass or clover. The crop should be sprayed when the leaf sheaths start to lengthen, but before the second node is detectable. The wild-oats should be growing vigorously.

Difenzoquat is a quaternary ammonium salt related to paraquat and diquat, but in contrast with the non-selective rapid scorch of these 2 compounds, difenzoquat is specific for the control of oat species. The symptoms develop slowly as chlorotic patches on the leaves followed by growth inhibition and necrosis. Difenzoquat can be used in barley, rye and specified varieties of durum, winter and spring wheat and triticale. The crop stage is not critical, but it is preferable to spray from early tillering to the one node stage at the full rate, but may be applied from the 3 leaf stage, if the half-rate autumn treatment is used. The wild-oats should have at least 2 leaves, but not be past the end of tillering.

Diclofop-methyl is the first member of a new group of herbicides, (esters of 2,4-(aryloxy)phenoxyalkanoic acids), effective on certain grass weeds. Diclofop-methyl is recommended for autumn and spring applications for the control of wild-oat, black-grass, rye-grass, rough meadow-grass *(Poa trivialis)*, and Yorkshire-fog *(Holcus mollis)*, in wheat, durum, rye, triticale and barley. Excessive scorch of the crop has been noted from applications to spring crops in the north of England and Scotland, but recommendations are made for autumn applications to winter cereals in the north of Britain. Absorption of diclofop-methyl by both roots and shoots occurs, with severe inhibition of growth resulting. It is translocated readily within the plant, selectivity probably being due to detoxification within resistant species. The weeds may be treated from emergence to early tillering depending on species. Wheat is very tolerant at any crop stage, but winter barley should be sprayed between emergence and the first node detectable stage (GS31) while spring barley applications are limited to the period between 4 fully expanded leaves (GS14) and 2 tillers (GS22).

Glyphosate is an organophosphorus compound which is generally recognised as a total herbicide, but has achieved an important niche in weed control in wheat and barley. It may be used pre-harvest of wheat, barley and oats, provided the mean grain moisture content is not greater than 30 %, and the seed is not to be used for multiplication or malting for whisky. It can be used post-harvest after all cereal crops. The main use to date has been for the control of couch *(Elymus repens)*, which is not well-controlled otherwise in cereals and is difficult to control in most other crops that may occur in rotation. This pre-harvest treatment is being used to control other persistent perennial weeds that are still green at harvest, and to reduce harvesting problems due to other weeds that have escaped routine herbicide sprays.

In the north of England and Scotland, pre-harvest applications have been shown to be more active on perennial weeds than applications made after harvest. There is also a recommendation for the control of perennial grass weeds before sowing spring cereals, but at least 10 cm foliage must be available, and treatment is not expected to be as effective as later in the season.

A formulation with additional wetting agent is available for pre-sowing control of annual weeds.

Mixtures and sequences of herbicides

(Tank-mixing is discussed on page 93)

Sequences of herbicides may be necessary if ready formulated mixtures or tank-mixture recommendations are not available to broaden the weed spectrum sufficiently to accommodate all the weed problems in a field. Apart from broadening the weed spectrum controlled, sequences are also recommended to deal with certain difficult weeds which germinate over a long period, e.g. wild-oats *(Avena fatua)*, and barren brome *(Bromus sterilis)*.

In most cases of sequential post-emergence applications, it is sufficient to allow the first spray to dry on the foliage before applying the second, but there are exceptions which, if not noted beforehand and the label recommendations closely followed, can lead to severe crop damage. For instance, the wild-oat herbicide, flamprop-methyl, which cannot be tank-mixed with any broad-leaved herbicide, also cannot be used in sequence with any herbicide containing 2,3,6-TBA, and there must be an interval of between 5 and 21 days between applications of other broad-leaved herbicides according to type and weather.

Manufacturers' recommendations about mixtures and sequences should be closely followed to avoid crop damage and other problems.

Fungicides

Chemical groups and properties

Cereal diseases have many causes, and may result in considerable reduction in yield and quality. The losses vary with the nature of the pathogen, the timing and intensity of attack and environmental stresses.

As such a wide variety of pathogens affect cereals, the control measures need to be equally diverse. The effectiveness of these measures must depend on the pathogen, the tissue affected, the inherent biological activity of the fungicide and correct timing. Ultimately, the efficacy of any fungicide depends on the amount of active ingredient that arrives at the site of action.

Several biological properties are desirable in a fungicide for effectiveness: it must protect the host from infection, suppress the colonisation of injured tissues, and destroy the inoculum available for new infections. Most fungicides used commercially for cereal disease control are effective as protectants, but relatively few have additional curative, eradicant and antisporulant activity.

Fungicides may be grouped according to their mode of action and method of treatment. They can be broadly defined as non-penetrant, surface protectant or penetrative fungicides. The last group may be further classified into those with localised penetration and those which are translocated within the host, away from the initial source of introduction; the "systemic" fungicides.

Non-penetrant fungicides should have the ability to kill pathogens on contact, suppress growth of superficial mycelium and/or limit spore generation. They are usually most effective when applied before infection occurs, as protectants. Biological effectiveness in non-penetrant fungicides depends, therefore, on some degree of persistence, and often they have to withstand much weathering as surface deposits. A factor governing their efficiency and biological activity is the re-distribution pattern of the active ingredient on the host plant. Often rain-fastness is considered an important attribute of a non-penetrant fungicide, but some re-distribution of surface deposits by rain splash as solutions or suspensions, or by volatilisation from the surface of the crop, may enhance the distribution pattern and affect the active ingredient.

Penetrative fungicides, on the other hand, aim to give better biological performance through their ability to attack the internal hyphae, mycelium and haustoria of plant pathogens. They also tend to give improved distribution of active ingredient within the host plant.

Uptake and distribution of a systemic fungicide depend on whether it is applied to aerial or subterranean parts of the host plant. Movement of the majority of systemic fungicides is mainly upward in the transpiration stream and much dependent on the particular host species, the physiological status of tissues and the environmental conditions during and after treatment. Entry of fungicides applied to leaves is very dependent on the cuticle and associated structures. This entry is followed by movement across leaves and towards leaf tips and margins.

Both non-penetrant and penetrative fungicides have been used for seed treatments. The non-penetrant fungicides, e.g. organomercury, destroy pathogenic organisms on the seed-coat and sometimes afford protection to the germinating seedlings from soil-borne inoculum. Penetrative fungicides, e.g. triadimenol, are used against fungi that invade the embryo, which is inaccessible to contact fungicides, and provide a reservoir of active ingredient during crop growth for the control of air-borne diseases.

Because of the risk of disease organisms developing resistance to the chemicals used to control them, it is convenient to classify fungicides in groups according to their differing biochemical modes of action (see Table 4.7). The risk of fungicide resistance developing when using chemical control systems is reduced by using multi-site fungicides where ever possible; avoiding the repeated application to a crop of the same fungicides or of different fungicides which have the same mode of action; and by using formulated fungicide mixtures or label recommended tank-mixes of fungicides in which more than one of the fungicides is active against the same disease or diseases involved.

Penetrative fungicides

Benzimidazoles are broad spectrum, systemic fungicides. They are used principally on cereals as foliar sprays against eyespot, but seed and soil treatment on other crops is not uncommon. At low concentrations, benzimidazoles prevent elongation of germ tubes of sensitive fungi; at higher concentrations, they reduce spore germination.

Triazoles and **imidazoles** are also broad spectrum systemic fungicides. They are used principally for control of mildew and most foliar diseases of cereals. They act by interfering with ergosterol biosynthesis. Some are used as seed treatments.

Pyrimidines are very specific in their action and are used to control powdery mildew, formulated as foliar sprays or seed treatments. They are active at low concentrations and appear to interfere with purine metabolism. They are systemic and readily absorbed by roots, then translocated upwards to leaves. They can also inhibit mildew spore germination prior to penetration, and also act at later stages.

Morpholines are active against mildew and rusts. Their precise mode of action remains uncertain, but recent evidence indicates that they interfere with ergosterol biosynthesis at a different stage from that inhibited by triazoles. They are used as foliar sprays.

Carboxamides are used almost entirely on cereals as a seed treatment to control smut and bunt. Their penetrant action allows the control of loose smut which is situated deep in the seed.

Pyrazaphos is used mainly for control of powdery mildew as a foliar spray. The uptake by roots following seed treatment is inadequate for systemic control. It is important because its mode of action differs from those of other fungicides which are encountering disease resistance problems.

Non-penetrant fungicides

These principally have a protective, non-systemic action, although some may eradicate surface infections. They are mostly toxic to fungal spores and are multi-site inhibitors, affecting several different enzymes.

Fungicide mixtures and development of resistance

Frequent applications of fungicides are not likely to be economically worthwhile in most cereal crops, and are considered undesirable both environmentally and because they may increase the risk of fungicide resistance.

The risk of resistance developing is greater with fungicides which act at a single site, or a limited number of sites in the fungal cell, than with multi-site fungicides. Usually fungi resistant to one fungicide are resistant to related fungicides in the same group, with the same mode of action. In cereals, fungicide resistance has caused relatively few problems in the past. Following the introduction of ethirimol to control barley mildew, isolates with decreased sensitivity were detected. The development of less sensitive isolates decreased the persistence of control. More recently, decreased sensitivity has been recorded with triazoles, but not clearly associated with poor control. However, isolates of the eyespot fungus resistant to the benzimidazoles have been associated with control failure in some cereal crops, and resistant forms detected in most cereal growing areas of the United Kingdom.

TABLE 4.7 Cereal fungicides: classification and use

Group number and collective name	*Examples*	*Major use in UK cereals*
(a) *Site specific fungicides – penetrative*		
1. Benzimidazoles	benomyl carbendazim fuberidazole thiophanate-methyl	broad spectrum: eyespot, mildew, *Septoria* and *Rhynchosporium*
2. Dimethylation inhibitors		
Triazoles	flutriafol propiconazole triadimefon triadimenol	broad spectrum: mildew, net blotch, *Rhynchosporium, Septoria*, and rusts
Imidazoles	prochloraz imazalil	Principally for seed treatment against seed-borne diseases and mildew.
Pyrimidines	nuarimol	
Piberazines	triforine	
3. Morpholines	fenpropimorph tridemorph	mildew and rusts
4. Hydroxy-pyrimidines	ethirimol	seed treatment and sprays for mildew.
5. Carboxyamides	carboxin	seed treatment: smut, bunt
6. Dicarboxyamides	iprodione	*Septoria*
7. Guanidines	guazatine	seed treatment
8. Organophosphates	pyrazaphos	mildew (net blotch, *Rhynchosporium*)
(b) *Multi-site fungicides – non-penetrant*		
9. Phthalimides	captafol	*Rhynchosporium, Septoria*
10. Phthalonitriles	chlorothalonil	*Septoria*
11. Dithiocarbamates	maneb mancozeb propineb thiram zineb	In mixtures with fungicides from other groups, broad-spectrum
12. Mercurials	phenyl mercury acetate	seed treatment
13. Sulphur	sulphur	mildew

Fungicides are usually more effective in controlling low amounts of disease, where there is less likelihood of resistance developing. However, routine prophylactic sprays of the same chemical or chemicals from the same group with the same mode of action should be avoided. If a sequence of sprays is required for effective disease control, fungicides should be selected from groups with a different mode of action for subsequent treatment and full doses used. There is currently frequent reference to the repeated low dose, or "little and often" approach, which is sometimes expected to give better disease control with lower costs. There is a possible danger from this type of approach, where pathogens may be subjected to sub-lethal doses of active ingredient for extended periods, thereby increasing the risk of resistance developing. Recent field experiments by ADAS have shown that the repeated low-dose method of protecting crops against disease has no advantage over 2 well-timed sprays. Also, if energy and machinery costs are included with the cost of fungicides, the "little and often" approach is frequently not as cost-effective.

Combinations of fungicides sprayed together can give several different responses. This can include antagonism, synergism and additive effects. Concurrent with these responses, changes can occur in phytotoxicity and in the spectrum of affected species.

Many fungicide mixtures are commercially available and frequently used in cereals for disease control. These have been tested, formulated and approved for use. Appropriate mixtures can decrease the risk of resistance developing, providing each constituent is effective, and that they include fungicides from both site specific and multi-site groups. Tank mixes of fungicides should only be made following manufacturers' recommendations because the active ingredients are often formulated differently, and the spray supplements such as wetters and spreaders may not give the desired performance when combined.

Insecticides

Chemical groups and properties

There are 21 chemicals presently approved for the suppression of pests of cereals in British agriculture; two are molluscicides and the remainder are insecticides.

Molluscs and insects may affect the establishment of the crop (e.g. slugs, wireworm and leatherjackets); affect the growing plant, (e.g. cereal aphids, wheat bulb fly, wheat blossom midge); or transmit viral disease to the growing plant (e.g. BYDV transmission by cereal aphids).

The control measures for this variety of pests need to be very diverse, and the chemicals may act as appropriate control measures for different pests under different conditions, dependent on timing of application and placement.

Molluscicides

Two molluscicides are used in cereals: **methiocarb** and **metaldehyde**. Methiocarb kills slugs by contact action, and is a stomach poison when ingested, apparently acting on the central nervous system. Metaldehyde paralyses slugs, which, in dry and windy conditions, desiccate and die. These molluscicides are applied overall through fertilizer distributors and seed drills, from the air or by hand, and are formulated in barley meal or bran pellets.

As slugs damage the emerging and establishing plant, such treatments are applied early, before or as soon as symptoms of attack are seen.

Organochlorines

Gamma-HCH is an organochlorine used to control leatherjackets and wireworms. It is formulated as a bran-based pellet, a seed treatment and a suspension concentrate. Gamma-HCH is a stomach poison, and has fumigant properties. It affects the insect central nervous system.

Organophosphates

Of the organophosphorus insecticides presently approved for cereal pest control, 6 are systemic; these are:

demeton-s-methyl;
dimethoate;
heptenophos;
omethoate;
oxydemeton-methyl;
thiometon.

They are all formulated as emulsifiable concentrates for use as sprays against cereal aphids. Omethoate is used to control wheat bulb fly and frit-fly in spring-sown cereals and cereal flies in winter wheat and barley. Dimethoate also controls wheat bulb fly.

The non-systemic insecticides are:

carbophenothion;
chlorfenvinphos;
chlorpyrifos;
fenitrothion;
fonofos;
phosalone;
pirimiphos-methyl;
quinalphos;
triazophos.

They are recommended for the control of a variety of cereal pests, including wheat bulb fly, aphids, thrips and yellow cereal fly. Most are supplied as normal formulations for application as sprays, but some targets require specialised formulation:

wheat bulb fly
- liquid seed treatment (e.g. carbophenothion)

- sprays (e.g. triazophos)
- granules (e.g. fonofos)

frit-fly
- sprays (e.g. chlorpyrifos)

wheat blossom midge
- sprays (e.g. fenitrothion)

aphids
- sprays (e.g. phosalone)

thrips
- sprays (e.g. fenitrothion)

yellow cereal fly
- sprays (e.g. triazophos)

The systemic organophosphorus insecticides are stomach poisons. Some of the systemic materials have fumigant action or translaminar activity. They are all neurotoxins, the symptoms of effect being restlessness, hyperexcitability, tremors, convulsions and paralysis, resulting eventually in death.

Carbamates

Pirimicarb has contact, fumigant and translaminar actions, and is relatively safe to non-target insects like bees and ladybirds. It is a neurotoxin, resulting in similar symptoms to organophosphorus poisoning, and is recommended for the control of cereal aphids.

Synthetic pyrethroids

Synthetic pyrethroids are recommended for use in the autumn to control the aphid vectors of BYDV. They have both contact and stomach action, and are very fast-acting. They affect both peripheral and central nervous systems of insects. Permethrin was the first agriculturally important synthetic pyrethroid, developed in the early 1970s, but it is being superseded by a more active group, including cypermethrin and deltamethrin.

Cereal Growth Regulators

Cereal growth regulators are mostly used to shorten and strengthen straw to reduce pre-harvest lodging. They may also reduce head-loss by increasing straw thickness, and there is some evidence that improved rooting, increased tiller number and increased grain number per ear may also be possible in certain situations resulting in increased yields under dry conditions.

There are three growth regulator products available for use in cereals: chlormequat, 2-chloroethylphosphonic acid, and mepiquat chloride + 2-chloroethylphosphonic acid.

Chlormequat is thought to inhibit giberellin synthesis, so preventing cell elongation. It is taken up through leaves and roots, but transport within barley is much reduced compared with winter wheat and oats. As a consequence, the straw-shortening effect is usually maintained through to harvest in wheat and oats, but not in barley. Over 10% reduction in height has been obtained in tall-strawed winter wheat, but straw length is generally reduced by 1% or less in winter barley. Short-strawed wheat varieties have, however, shown little or no yield response. Lodging prevention treatments are applied as stem elongation starts (GS30-31). Recommendations are made for use in spring-sown crops of wheat and oats, and all winter-sown cereals.

In winter-sown barley, chlormequat has been used in an attempt to modify development and ear uniformity. Yield responses in trials have been similar to those for wheat on lodging sites. However, on non-lodging sites, response has generally been lower and more erratic with some yield reduction.

2-chloroethylphosphonic acid releases ethylene gases within the cereal plant after entering mainly through the leaves. The exact mechanism of ethylene effect on cell extension is not known, but may well be through inhibition of auxin activity.

A reduction in straw length in the order of 8-12% is common in winter barley, and similar results are claimed on winter wheat. Lodging in winter barley is typically reduced from 50% in untreated plots to 10-15% where 2-chlorethylphosphonic acid has been applied, but the greatest benefit has often been in delaying the onset of lodging. Effects on winter wheat have not been as extensively tested, but better lodging prevention than chlormequat is claimed where lodging pressure is high. Yield responses on winter barley have ranged from 4-15% increase on sites where lodging has occurred, but have been small or negative on non-lodging sites. Commercial recommendations are now made for use in triticale and winter rye.

Mepiquat chloride is sold in mixture with 2-chloroethylphosphonic acid. It has similar activity in barley to chlormequat, that is the prevention of synthesis of giberellins and hence cell extension. It is claimed to be more readily translocated than chlormequat in barley. Reductions in straw length and lodging by the mixture are similar to 2-chloroethylphosphonic acid alone. Mepiquat chloride is also claimed to increase straw thickness. Yield responses in trials are similar to 2-chloroethylphosphonic acid alone on lodging and non-lodging sites. The exception is the winter barley cv. Sonja, where economic responses have been obtained on both lodging and non-lodging sites. Yield components, however, have rarely shown significant responses. The mixture is recommended for use in winter barley, winter rye and triticale from the 2 node detectable stage (GS32), to flag leaf just visible stage (GS37). Recommendations are also made for the use of the mixture in specified barley varieties in Scotland and Northern Ireland only.

Mixtures of Agrochemicals

The application of tank-mixes of crop protection chemicals recently became a principal focus of attention on all sides of the industry as the simultaneous application of agrochemicals assumed an important part of cereal growing in order to reduce labour and save time. Unfortunately, the increased use of mixtures was not accompanied by a comparable advance in knowledge concerning physical, chemical and biological compatibilities of such mixtures.

The British Crop Protection Council, (BCPC), included a consideration of this problem as a feature in a number of their Annual Reviews of the Usage of Herbicides, Insecticides, and Fungicides, starting in 1977. In 1979, they initiated a study of the use of tank-mixes by two hundred cereal growers throughout the UK. This study revealed that the vast majority of tank-mixes used by these farmers was not supported by any official manufacturers' recommendations. In order to obtain a comprehensive picture of the likely effects of mixing agrochemicals on cereal crops, the products must be tested extensively in the laboratory and field, and without sufficient evidence of physical, chemical, toxicological and biological compatibility, manufacturers are hesitant about making tank-mix recommendations. Furthermore, many Acts of Parliament, in particular, the Health and Safety at Work Acts and Trade Descriptions Act, contain certain provisions which could make the manufacturers liable should damage occur, following the use of tank-mixes which they have supported. Consequently, the range of supported mixtures is relatively limited compared with the number of products available on the market.

Nevertheless, the list of mixtures supported by manufacturers is continually being extended, and since 1980, an annual *Tank-mix Spray Guide* has been published by the Farmers Weekly journal, (Business Press International, Sutton), with the co-operation of the British Agrochemicals Association and guidance from the British Crop Protection Council. This Guide is a simply designed source of reference on label recommendations that manufacturers make for mixing their product with others.

If a desired tank-mix is not mentioned in the Guide or on a label, the guidance of the manufacturers of the components should be sought.

The manufacturers may in many cases have evidence that a mixture is physically compatible in the tank; such a test is simple to carry out, but this test will not give any information about chemical or biological activity, or effects on the toxicity of the components. The differing formulations of mixture components may be designed to get the spray to quite different targets, and cases of crop damage from the mixing of apparently physically compatible products are common, as are cases of reduced efficacy of one or more of the components.

The addition of adjuvants such as wetting agents complicates the situation further, and the safety of the crop and the user may be jeopardised by some materials.

Mixing with other additives, in particular trace elements, should very carefully follow manufacturers' recommendations, as chemical reactions of such mixtures may lead to the precipitation of insoluble salts, which can block the sprayer and settle in the tank. In particular, the use of unformulated manganese sulphate in mixtures with agrochemicals has been a common source of problems.

The addition of liquid fertiliser commonly causes or increases crop scorch or damage, and such a practice is not generally recommended.

Other major problems related to tank-mixes are the differing spray volumes and pressures that may be required for different components of a mixture, and the differing crop growth stage timings that may be recommended. If the components of a mixture have different crop growth stage application recommendations, then the most restrictive recommendation must be followed. The herbicide recommendation is usually the most restrictive one. It is important, if a fungal infection warrants immediate treatment, not to wait for the herbicide timing so that a mixture can be safely applied, but to apply the herbicide and fungicide separately. Not to do so could incur yield losses which could be greater than the extra cost of spraying the components separately.

TABLE 4.8 Tank-mixing considerations

Advantages:
saves time/labour
reduces energy costs
reduces passes through the crop
increases the chance of applying all sprays when suitable spray occasions are few
Disadvantages:
not much known about physical/chemical compatibility
biological efficacy of components may be affected
crop growth stage may not be correct for all components of the mixture
may be too early or late for best effect on one of the intended targets

Sequential Applications of Agrochemicals

Sequential application of different agrochemicals is an even more complicated subject than tank-mixes, because there are even more field variables to take into account. Most manufacturers state whether an interval is necessary between the application of their product and certain other products, but no comprehensive list of "safe" sequential treatments exists.

The best course for farmers is that they should study the manufacturers' recommendations for both components of the proposed sequence. If there is no caveat regarding either the sequence of chemical use, or the time interval which should elapse between applications, then there should be plenty of flexibility in the timing of the second application. If there is any doubt, the product manufacturers should be approached for advice.

Chapter 5 THE APPLICATION OF CROP PROTECTION CHEMICALS

General Considerations

The application of crop protection chemicals to cereals requires a consideration of both the target and the means of applying what is often only a very small amount of active ingredient over a large area. The active chemical is normally formulated with other materials as described in Chapter 4, to facilitate its even distribution.

Some seed treatments are available as dusts or undiluted liquids which are mainly applied with specialised equipment by seed merchants. A few herbicides are sold as granules requiring the use of special granule distributors, and molluscicides as pellets which can be applied by spinning discs or other fertiliser distributors; but the majority of crop protection products for cereals are formulated as aqueous concentrates, emulsifiable concentrates, wettable powders, water-dispersible grains or suspension concentrates ('flowables') which require dilution with water before application through a spraying system.

The principles underlying the application of spray treatments are similar for all crop protection chemicals. The product must reach the target, be it soil, crop or weed foliage, insect exoskeleton or fungal structures on the foliage, and remain on the surface long enough to take effect. In most cases some penetration of the surface of the target is required, and in many cases further movement of the chemical within a plant or animal is necessary to achieve the desired response.

Methods of application and formulation type interact in complex ways, varying the response of target organisms to active chemicals. Considering the long period during which crop spraying has developed surprisingly little is known about the precise requirements for the most efficient application of the majority of crop protection chemicals.

Most spraying systems are a compromise between a number of factors such as the need to have a sufficient number of small drops to obtain the evenness of cover required and the likelihood of small drops causing problems by drifting off-target; or the need to have sufficient water to produce the desired number of larger drops yet not to require so much that the logistics of spraying are made difficult through the size of spray tank necessary and water supply problems.

The targets

The initial targets for crop protection chemicals used in cereals are mostly crop or weed leaves, or the soil in which the crop is growing, or will be growing. Less frequently the target may be insects or pathogens present on the crop. These are the targets for prime consideration in selecting a spraying system. However it must be remembered that the ultimate target may be an insect or disease organism that is expected to arrive on the crop later, or weeds that are expected to germinate or emerge through the soil some time after the treatment is applied.

Foliage application of crop protection chemicals

Contact herbicides have a zone of action restricted to the area of spray contact, or immediately adjacent to the spray drop. Translocated herbicides move through the plant and may only need to be retained on a part of the target plant, so evenness of foliage cover is of less importance than for 'contact' herbicides.

The size of the weeds affects the desirable characteristics of a foliage treatment; larger plants require a better cover of contact herbicides to obtain a complete kill than small plants. The contact herbicides most commonly used in cereals, ioxynil and bromoxynil, are found in practice to be much more active on small young weeds than on large plants.

The shape and leaf orientation of plants affects the ability of chemicals to be retained on the foliage. Cereal crops, with their upright leaf orientation, tend to retain less of the spray than broad-leaved weeds and this is an important factor in determining the selectivity of some cereal herbicides. The upright nature of most young grass weeds makes them difficult targets for foliage-applied herbicides.

The nature and distribution of wax on the surface of the leaves is an important factor in determining the rentention of spray deposits. For example, the mealy surface of fat hen *(Chenopodium album)* produced by wax globules, retains water far more readily than the smooth wax surface of field pansy *(Viola arvensis)*. This is reflected in the ease of control with foliage-applied herbicides of fat hen compared with field pansy. Various environmental factors, in particular temperature, influence the formation of wax and subsequently the effectiveness of foliage-applied herbicides. Young rapidly growing plants have poorer wax development, thus more readily retaining and allowing the penetration of spray chemicals.

Hairiness of leaves may also affect retention of the spray; in general, although not in all cases, very hairy leaves are more difficult to wet than those with less dense hairs.

Spray falling on leaves that remain wet from rain or dew may run-off or be diluted beyond the point of effectiveness. Low volume sprays are less likely to be affected in this way than the more conventional spray volumes of 200 l/ha or more.

As in the case of herbicides, the application of fungicides and insecticides which work by direct contact with target organisms such as fungal structures on the foliage and insects with limited mobility (e.g. aphids), require greater care in distribution for maximum cover than those which work systemically through plant tissues, or are translaminar moving across leaf tissues to affect pests hidden from the spray by the angle of the leaf, or where pests are sufficiently mobile to make contact with deposited chemical (e.g. midges).

An improvement in spray cover can usually be obtained by using a spray system giving an increased number of smaller drops which, under pressure, can penetrate the crop canopy more efficiently. Chemicals showing systemic or translaminar activity can be applied using larger drops and lower spray pressures.

Growth regulators require application and formulation in much the same manner as translocated herbicides, foliage-applied systemic fungicides and insecticides. Evenness of application is not required to be as precise as for contact-acting chemicals. Nevertheless, good adhesion to crop foliage is required, and formulations take this and the need for leaf surface penetration, into account.

Soil applications

Some herbicides and insecticides may be applied direct to the soil, or in the case of some insecticides, in the seed drill. Cereal fungicides and growth regulators are not applied directly to the soil, but some fungicidal seed dressings may be taken up from the soil around treated seed *via* shoots and roots.

Overall applications of soil-applied herbicides for cereals are required to form a layer of chemically treated soil through which weeds will not grow when uptake is mainly through shoots, or to reach the root zone of the weeds when root uptake is important. This requires an even distribution of the herbicide on a reasonably fine tilth, so that optimum contact with the shoots and roots of both germinating and emerging weeds may be achieved. Many soil-applied herbicides require redistribution within the upper soil layers to increase the chance of uptake by the weeds. This redistribution can be achieved mechanically by the use of soil tillage machines or by natural means such as leaching under the influence of rainfall, diffusion as a vapour, or by moving *en masse*; the size of spray drops and evenness of spray are not highly critical so long as the intended dose of chemical is applied. Volatilisation of certain chemicals away from the soil can be a problem, unless the chemical is incorporated. The volatilisation of the grass weed herbicide tri-allate has been partially overcome by a granular formulation which does not necessarily need to be incorporated. The active ingredient is released from the granules by moisture; under dry conditions incorporation may be needed to achieve reliability. Even distribution of these granules is as important as even spray distribution since the vapour effect is limited to a very small area around each granule.

Little can be done to reduce the leaching of herbicides which have a relatively high water solubility. In conditions of very heavy rainfall, particularly in very sandy or gravelly soils, leaching in solution or mass movement of relatively insoluble herbicides may also occur. The selectivity of some of these soil-applied herbicides may be partly determined by the extent of movement of the herbicide in the soil in relation to the sowing depth and subsequent rooting depth of the crop seed. Abnormal weather patterns following the use of soil-acting herbicides can from time to time cause severe damage to the crop and reduce control of surface germinating weeds. Soil-acting herbicides are not recommended for use in soil conditions where the recommended sowing depth cannot be achieved.

In most soil types, and under normal rainfall conditions, there is relatively little leaching of the commonly used soil applied herbicides in cereals. Formulation, apart from granulation, probably makes little difference to the distribution of the chemical within the soil, although some experiments have shown that very high levels of wetting agent may increase the activity of some root absorbed herbicides.

The organic matter and clay content of the soil are the dominant factors determining the mobility and accessibility of soil-applied herbicides, because of the tendency of such particles to both absorb and adsorb chemicals. Product recommendations sometimes set limits to the level of organic matter and clay content in the soil where chemicals are to be used.

Cloddiness of tilth or the presence of many stones, particularly limestone brash, prevents the successful use of soil-applied chemicals because it is difficult to achieve an even distribution in the soil.

Soil-applied insecticides are also available as granular and spray formulations. The granules may be applied overall, as in the case of slug pellets, and left on the surface, or, as in the case of fonofos for wire-worm control, require incorporation into the zone of pest activity. The mobility of the pest reduces the need for absolute precision of distribution, but it has to be sufficiently even to make the chances of the pest coming into contact a certainty. Sprays can be used to control soil pests, and in some cases they may require incorporation to form a deeper zone of activity. Later sprays for egg-hatches, e.g. yellow cereal fly in the autumn, and some at-drilling sprays may not need incorporation, but evenness of application over the soil surface then becomes more important.

As in the case of herbicides, apart from those produced for application as granules, formulation probably makes little difference to the distribution of chemical within the soil. Experiments suggest that coating very small particles of the active ingredient with polymers (microencapsulation) may assist delayed release of active chemicals into the soil environment, and such formulations could be sprayed with a liquid carrier, but there are no products available as yet using this principle to effect activity in the soil.

Chemical seed treatments

Seed treatments of insecticides and fungicides require formulations that ensure adherence to the seed coat, and in some cases allow penetration into the seed coat of the developing seedling structures. Such treatments may be formulated as dusts, in which case little should remain free to present a dust hazard to the user; liquid formulations are less likely to be hazardous to the user in the field. However, liquid insecticidal treatments must only be applied in officially approved premises. Seed treatments are generally unaffected by rainfall, soil type, soil condition, and other environmental factors providing these are within the limits for normal crop growth. Treated cereal seed does not have an extensive safe life and once treated, seed should be sown as soon as possible.

There are no chemical seed treatments currently used to aid cereal weed control or crop growth regulation in the UK.

Application Equipment

The initial quality of distribution of the agrochemical, whether soil, seed or foliage applied, depends on the effectiveness of the applicator. Equipment available includes sprayers, granule applicators and seed treatment machinery.

Sprayers

There are basically two types of spray system used to distribute agrochemicals on cereal crops; conventional hydraulic nozzles and rotary atomizers.

Spray systems should deliver the active compound efficiently, and in optimum quantities, to the intended target, without unduly contaminating non-target organisms or the surrounding area. Their efficiency of doing so, and the degree of efficiency that is required for optimum effect, is currently a matter of some discussion and research.

For an overall application such as a soil-applied herbicide the target is clear and the requirements of the spray system used are not highly critical. So long as the chemical is deposited on the ground evenly without too much of a gap between spray drops, and the spray is restricted to the treatment zone of action, then the critical characterisitic of the application system should be that it applies the required dose. Drop size, range of sizes and concentration matters less than overall chemical rate requirement and soil condition.

For foliage applications the picture is more complex. The target for foliage herbicides is the weed, and not the crop (some herbicides are both foliage and soil acting). Fungicides, insecticides and crop growth regulator foliage applications are aimed at the crop, or the pest on or around the crop. The weed, the crop and the pest require sprays with different characteristics and in many cases the application system will be a compromise between conflicting ideals.

Conventional hydraulic systems

Most conventional hydraulic spray systems used by farmers consist of a tank of 200-1200 l capacity, a pump driven directly by the power take-off from the tractor, and a horizontal boom usually within the range 6-13 m, but increasingly of an even greater length.

The nozzles are most commonly of fan or cone type. The fan nozzle produces a flat fan-shaped spray whilst the cone nozzle produces a hollow cone of spray. The fan pattern and drop size are varied by changes in the spray angle of the emerging sheet, hydraulic pressure and orifice size. The drop size from cone nozzles can be varied by a change of swirl port and pressure.

Herbicides, systemic fungicides and insecticides and growth regulators are generally applied at pressures of 2-3 bars through fan-nozzles, at volumes of 80-200 l/ha. At these pressures and volumes there is a tendency towards producing a majority of drop sizes which do not create drift problems. However, a full range of drops from 1-100 μm in diameter are produced, including a very large number of small drops of less than 100 μm which are liable to drift over long distances. Although the total amount of active ingredient falling outside the target area may be very small, the environmental effects may be very undesirable. The very large drop fraction, on the other hand, may be ineffective due to run-off from the foliage leading to a relatively small proportion of the spray drops being retained by the foliage. It is highly likely that the range of drop sizes produced is a reason for hydraulic spray systems giving generally reliable results over a wide range of conditions.

Fine drop sprays are generally recommended for contact acting fungicides and insecticides. These can be achieved by a range of combinations of pressure and cone nozzle size. This practice, however increases the fraction of the spray liable to drift. There are a few instances when herbicides are more effective at low volumes and high pressures, but in general this is not the case, and this practice should be avoided. Clear guidance on how to apply chemicals effectively and safely is given in product labels.

Differences in drop size and volume requirements for various uses of agrochemicals must be borne in mind when mixing is contemplated.

Nozzle size and type are defined by different manufacturers in many different ways which tend to confuse users. A simple scheme using familiar words to define the quality of the spray has been proposed and is becoming accepted by machinery manufacturers, chemical manufacturers and operators. This is shown in Table 5.1. ADAS now provides listings of the nozzles produced by major manufacturers which fall into the various categories of coarse, medium, fine and very fine. There is a further category of very coarse which is used for liquid fertilisers rather than pesticides. The listings of nozzles are available from ADAS both in written form and on Prestel.

TABLE 5.1 Spray characteristics of proposed standard nozzle type groups

Type of nozzle	Coarse	Medium	Fine	Very fine
Characteristics of spray	Mostly coarse few fine drops	Wide range all drop size	Mostly fine few large drops	Mostly fine very few large drops
Retention on leaf surface	Poor	Moderate	Good	Very good
Potential drift hazard	Negligible	Low	High	Very high
Typical target or pesticide use	Soil-acting herbicides	Broad-leaf weeds, mixed species	Fungicides, insectides, grass weeds	Fungicides, insecticides

Rotary atomizer systems

Rotary atomizers are used to produce sprays with a more uniform drop size than hydraulic sprayers. The size of the drop is primarily related to the speed of rotation of a disc or cage, and the flow rate. The technique has been developed to allow low volumes of crop protection chemicals to be applied without recourse to the use of small driftable drops, or fine nozzle orifices that could block.

Whilst there is some evidence suggesting increases in efficiency of transfer of the agrochemical to the target surface, other factors such as concentration of the agrochemical and drop dispersion may reduce the initial benefit. There is little conclusive experimental evidence to back the suggestion, which in theory would seem tenable, that doses of crop protection chemicals can be reduced from those used in the conventional systems. Many trials are being undertaken by official bodies and agrochemical companies with these systems. The evidence indicates that herbicide efficacy has shown no improvement compared with efficient hydraulic systems and those depending on contact tend to perform less well. There is probably more scope for this type of application with fungicides and insecticides when smaller drops can be used without the same degree of environmental hazard that might be caused by herbicide drift. In practice, however, the response is patchy and unreliable, possibly due to insufficient penetration of the crop canopy by the fine drops. The problem of penetrating the crop canopy is attracting further development, for instance by the use of partially shrouded vertically mounted spinning heads and by directing air-streams containing the drops into the crop.

It should be noted that at present only a limited number of herbicides, fungicides and insecticides are cleared for commercial use in such systems.

At present those herbicides cleared for use are applied at 10-45 l/ha and insecticides and fungicides at 1-5 l/ha total spray. The low spray volumes used are one of the major advantages of these spray systems. Manufacturers' instructions should be followed as to the spray volume used and which of the several systems available can be used. Only those products with specific label recommendations for the system proposed should be used.

Control of drop size does not ensure that drift will not occur. Herbicides in particular should never be applied at a drop size less than 100/μm, and even at recommended sizes of about 170-250/μm, drift can still occur in windy conditions (5 mph).

The use of small drops with low amounts of diluent through rotary atomizers can lead to a concentration of the chemical which greatly alters the efficacy of the chemical and its safety to the crop and user. This is particularly the case in hot weather when further evaporation of water-based or volatile oil-based dilutents can occur, increasing the concentration of chemical in the drop.

Aerial spraying systems

The basic spraying equipment used on aircraft is no different in principle from that used on ground equipment but it is engineered to a higher degree of sophistication. There are, however, dangers inherent in the application of agrochemicals by air. No hormone herbicides are recommended for application by aircraft in the UK because of the considerable potential for drift damage. Furthermore, use is restricted to agrochemicals effective at very low volumes. Only a small number of agrochemicals are cleared by aerial application to cereals in the UK. These are listed as an appendix in the Civil Aviation Authority's publication CAP414 *The Aerial Application Certificate* obtainable from CAA, Printing and Publication Service, Greville House, 37 Gratton Road, Cheltenham, Gloucestershire GL50 2BN. Guidance can also be obtained from the National Association of Agricultural Contractors.

The main advantages in the use of aircraft are the speed of application, and the ability to spray crops when soil conditions are poor. However, the technique is expensive, there is less direct control by the farmer over the spray operation and aerial spraying is a cause of great concern to the general public.

Granule applicators

Granular formulations of herbicides and insecticides are applied by special ground machine applicators. They consist of a hopper, a metering device and distribution tubes. In certain circumstances modified fertiliser distributors may be used, and some pesticide and herbicide granules can be applied with fertilisers. The mixing of fertiliser and granular formulations used in cereals is not satisfactory if there are very big differences in granule size which would result in sedimentation and uneven flow.

Seed treatment machinery

Treatment of seed with insecticide or fungicide requires a dusting or spraying machine and the process should take place in an enclosed space which the operator should not have to enter. Nevertheless, the operator must implicitly follow the manufacturer's instructions about the wearing of special protective clothing.

The chemical should have been formulated by the manufacturer so that it adheres well to the seed, decreasing any vapour or dust hazard to the operator.

Electrically charged droplet systems

Research and development is in progress with a number of systems of electrically charging spray droplets to improve spray distribution and deposition. These include modification of existing hydraulic nozzles, rotary atomizers and a novel nozzle to produce droplets electrostatically. Pesticides for the latter nozzle have to be specially formulated. Drop size can be controlled by changing the voltage and flow rate. The use of charged sprays offers possibilities of improving the efficiency of application so that a higher proportion of the spray reaches the intended target, including the underside of the leaf. In the future this may allow doses of some chemicals to be reduced and minimise effects on non-target species.

Application in the Field

Application of herbicides in the field is comprehensively covered in the British Crop Protection Council's *Weed Control Handbook: Principles*, Ed. H. Roberts, 7th Edn. 1982, and of insecticides and fungicides in the *Pest and Disease Control Handbook* Eds. Nigel Scopes and Michael Ledieu, 2nd Edn. 1983, BCPC, Croydon. General advice on safe use of crop protection chemicals is given in ACAS *Approved Products for Farmers and Growers* handbook 1985 (MAFF RB 380, HMSO). The Agricultural Training Board organises practical courses on crop spraying, details of which can be obtained from local ATB training advisers.

Low ground pressure vehicles

A recent development of particular interest for spraying cereals is the use of low ground pressure vehicles (LGPV). Low ground pressure is usually defined as a wheel exertion pressure of less than 0.35 kg/cm^2 (5 lb/in^2).

The interest has been stimulated by the move to earlier drilling and the increase in winter cereal cropping. This has increased the need for autumn, winter and early spring spray application when ground conditions are often unsuitable for spraying vehicles or conventional field-working tyres.

Low ground pressure has been achieved by developing self-propelled sprayers which have a large number of wheels, or tracks, or very large wheels and tyres, and also by adding double wheels with large tyres and special treads to tractors.

One aspect of low ground pressure spraying that is sometimes overlooked is the likelihood of poor conditions for pesticide activity during the winter months. Wet cold conditions can increase the sensitivity of the crop but decrease the control of the pest, pathogen or weed. For example the herbicide mecoprop applied under wet, cold conditons followed by night frosts can cause crop scorch and affect overall growth.

Research on the effects of such conditions on the efficacy and safety of agrochemicals is limited. Manufacturers' instructions as to the conditions suitable for effective use of products should be followed.

The availability of a low ground pressure vehicle will increase the number of days on which spraying is possible when the only restriction is the unsuitability of the soil for normal traction.

Tramlining

Restriction of wheelings for all spraying operations to pre-determined tramlines produced by blocking coulters in the seed-drill will reduce crop damage throughout the season, increase the accuracy of spraying, and avoid overlapping. Tramlining requires the standardisation of equipment to match the pre-determined wheeling spaces.

Damage to crops from incorrect application

Damage to crops from incorrect application may be due to both machine faults and operator errors.

Overdosing with an agrochemical may adversely affect the crop, especially when a chemical is used in a marginal situation, such as the use of certain residual herbicides in very light or gravelly soils. Overdosing may also leave residues in the soil which could be harmful to following crops.

Overlapping of the spray swaths and overspraying of the headland whilst turning are the most common causes of overdosing. Overdosing may also be due to miscalculation of the required concentration, or to spillage, spraying from a stationary sprayer, dribble from nozzles (the use of anti-drip devices can prevent this), faulty nozzles, and slow-down of the tractor through wheel-slip or travelling on difficult or uphill terrain. Inadequate mixing of the chemical and water is a frequent cause of overdosing at the start of spraying.

Damage to the crop being sprayed can also be caused by mistiming the application, an aspect which is considered further in Chapter 4 (p 67).

Cereal herbicide spray drift damage to neighbouring broad-leaved crops frequently occurs. Drift is the result of using too high a pressure, nozzles that are too fine, having the boom too high above the crop or spraying in windy conditions. The growth regulator or hormone herbicides are the most likely to cause problems because they are so active at small doses on rapidly growing or developing broad-leaved crops.

Damage to neighbouring broad-leaved crops has been demonstrated to result from vapour drift arising from ester formulations of hormone herbicides sprayed onto cereals. The problem is usually associated with hot, relatively tranquil conditions and can occur during a period of several days after spraying. The use of ester formulations of hormone herbicides should be avoided in cereal crops adjacent to sensitive broad-leaved crops. These include oilseed rape, peas, beans, sugar beet, fodder beet, swedes, turnips, spring cabbage, cauliflowers, Brussel sprouts, brassica plantbeds, tomatoes, peppers, fruit crops, vines, ornamentals and hops. There are other sensitive crops but they generally emerge after the period when hormone weed killers are applied to cereals.

A British Crop Protection Council working party has discussed the influence of weather on the efficiency and safety of pesticide application, paying particular attention to the drift of herbicides. Their findings are published as BCPC *Occasional Publication No 3*, Eds. J. G. Elliott and B. J. Wilson.

General guidelines for the application of pesticides can be found in the MAFF Booklet 2272 *Guidelines for applying crop protection chemicals: ground crop sprayers for agriculture*, produced jointly by MAFF and BCPC.

Chapter 6 CROP PROTECTION OPPORTUNITIES IN AUTUMN-SOWN CEREALS

This chapter dealing with cereals and the succeeding one covering spring cereals, are organised to give both cultural and chemical crop protection opportunites. These are arranged under broad growth stage periods for weed, disease, pest and lodging control together with some general considerations.The Chapter finishes with tabulated guides giving some of the considerations to be borne in mind during each of the main growth periods.

These chapters do not go into any depth on matters such as comparative weed tables for different herbicide products: they deal mainly with the broad outlines of what can be done about the many different crop protection problems at various times of the year. More detailed current summaries of pesticide recommendations and comparative tables can be obtained in annually up-dated publications such as the Farmers Weekly Supplements: *Cereals Spring Herbicides* (issued in the first week of February), *Cereals Autumn Herbicides* (end of August), and *Cereal Fungicides* (end of January), which are published in association with ADAS by Business Press International, Sutton. These contain similar material to that given in ADAS Booklets 2252 *Weed Control in Cereals: Spring and Summer;* 2253 *Weed Control in Cereals: Autumn and Winter;* and 2257 *Uses of Fungicides and Insecticides on Cereals.*

For the fine detail of recommendations the current product labels and guides must be carefully studied and the particular precautions followed. Under the new legislation misuse of crop protection chemicals becomes a criminal offence and the label an even more important document than hitherto.

WEED CONTROL

As discussed in earlier chapters husbandry has a profound effect on the weed situation that develops within a cereal growing system. The opportunities for using cultural methods are discussed in later paragraphs in this section. The following section deals with choosing a herbicide type or herbicide programme and factors influencing decisions on weed control in winter cereals.

Systems of herbicide use

Anticipation of the main weed problems helps towards successful control and is implicit in any decision to use a pre-drilling or pre-emergence herbicide. Grass weeds are more easily identified in cereals as seed heads in June than as seedlings in November especially if only low numbers are present. If crops have to be walked in June for disease inspection, there is an opportunity at this time to identify weed problems and upon this assessment decide which herbicides to use in the next crop. With broad-leaved weeds a knowledge of the weed history will be necessary to make decisions on pre-emergence control.

Pre-emergence weed removal avoids possible yield loss through weed competition. Post-emergence application of herbicides in the autumn to obtain a weed-free crop in the early spring will in most circumstances avoid damaging competition unless very high infestations of competitive species such as black-grass, brome or cleavers are present. Delaying control until late spring runs an increasing risk of yield loss through weed competition.

Many farmers on heavy land have in the past preferred to apply a pre-emergence herbicide as soon as possible after drilling to guarantee their weed control treatment before bad weather and soil conditions made application impossible.

Recent trends towards earlier drilling, and the use of low ground pressure spray vehicles, have now resulted in greater reliance on early post-emergence materials. Weed control from soil-applied herbicides tends to be more variable than from those applied to the weed foliage. Individual herbicides do vary in this respect. Many require moist soil to be fully effective, and performance can be reduced on organic soils. Some are reduced in performance by the presence of ash and straw residues built up with repeated non-plough cultivations. Some post-emergence herbicides which act both through the foilage and the soil may also be affected to some extent by the presence of ash and straw residues.

Post-emergence applications do allow seedling weeds present at the crop 2–3 leaf stage to be assessed, and decisions on whether or not to spray and which herbicide to choose to be made accordingly. More weeds may well emerge over the winter period but the earliest and most competitive fraction can be removed with a late autumn post-emergence spray. Control of species such as wild-oats,where emergence extends into the spring, will be more complete if delayed until late spring, but if infestations are sufficient to threaten crop yields a sequence of an autumn treatment followed by a spring treatment is recommended.

Herbicide choice will also be influenced by the association of weed species present. For example if wild-oats are the only target a specific wild-oat herbicide such as difenzoquat will suffice. If however both wild-oats and black-grass are present the choice will depend on the relative numbers of the two species. If predominantly wild-oats, then a herbicide such as flamprop-isopropyl should give good control of wild-oats and suppression of black-grass. Conversely a mainly black-grass situation will require a urea-based herbicide, primarily for black-grass but with some control of wild-oats and control of other grasses and some broad-leaved species as a bonus. If wild-oats and black-grass are absent, other annual grasses and some broad-leaved weeds can be controlled by lower rates of exisiting black-grass herbicides or other residual herbicides.

Cleavers, speedwells and pansy are notably resistant to most of these annual grass weed herbicides. The need to broaden the spectrum of control has led to the development of formulated mixtures containing mecoprop and to recommendations for tank mixes of mecoprop with existing products. In recent years there has been an upsurge in the use of the contact-acting hydroxybenzonitriles, ioxynil and bromoxynil, either alone or as formulated mixtures or tank mixes with mecoprop. There is some evidence that mixtures of urea herbicides and mecoprop can predispose cereal crops to frost injury.

The effectiveness and spectrum of weed control can also be improved by applying herbicides in sequence. Tri-allate, originally recommended as a single treatment for wild-oats and black-grass, is now recommended to precede many other grass or broad-leaved weed herbicides to improve the performance against wild-oats, black-grass and brome (see Table 6.1).

Recent trials have indicated that sequential applications at lower than usually recommended doses of broad-spectrum broad-leaved weed herbicides, such as hydroxybenzonitrile plus mecoprop mixtures, can give more consistent control of larger weeds than a single application at the recommended dose. When conditions are ideal for the first low-dose application the need for the succeeding one may be removed.

Many recommendations for foliage-applied herbicides stipulate that weeds are growing actively for optimum herbicide performance. A few such as diclofop-methyl and bifenox with mecoprop work well under cold conditions and are suitable for winter application.

Large over-wintered weeds present in the spring may require special consideration. Over-wintered chickweed and cleavers can be controlled by mecoprop in mixtures at the early jointing stage of the crop; control is likely to improve as weed growth becomes more active with rising temperatures. Large mayweed and corn marigold are readily controlled by the clopyralid component of formulated mixtures, and large chickweed and cleavers by fluroxypyr.

With earlier drillings and poor weather in the spring, spraying often has to be carried out beyond the first node detectable stage. In general products containing one or more of bromoxynil, ioxynil, benazolin, clopyralid and mecoprop are recommended for spraying at the second node detectable stage. Fluroxypyr alone is recommended for applications made up to the time of flag leaf emergence.

Many products have recommendations for tank mixing with other pesticides, giving opportunities for integrated control and saving application costs.

TABLE 6.1 Herbicides recommended for sequential autumn or winter application after tri-allate in winter wheat and winter barley

		Type of weeds controlled in addition to wild-oats			
Active ingredients	*Time of use*	Black-grass	Meadow-grass	Brome grass	Broad-leaved weeds
chlorsulfuron + methabenzthiazuron	pre-[1] and post-[1] emergence	√	√		√
chlorsulfuron + metsulfuron-methyl	pre-[1] and post-emergence	(√)	(√)		√
chlortoluron	pre- and post-emergence	√	√	√	√
diclofop-methyl	post-emergence	√	√[2]		No
diclofop-methyl + ioxynil + bromoxynil	post-emergence	√	√		√
isoproturon	pre- and post-emergence	√	√	√	√
isoproturon + ioxynil + bromoxynil ± mecoprop	post-emergence	√	√		√
isoproturon + ioxynil + mecoprop	post-emergence	√	√	√	√
linuron + bifenox	pre-emergence[1]	(√)	√		√
methabenzthiazuron	pre- and post-emergence	√	√		√
metoxuron ± simazine	post-emergence	√	√	√	√
terbutryne	pre-emergence	√	√		√
trifluralin + linuron	pre-emergence	√	√		√

1. winter wheat only
2. rough meadow-grass only

(√) indicates partial control

These are some of the main considerations in selecting an appropriate herbicide. Details of individual herbicides are listed in the following table. Where several herbicides appear appropriate to the specific crop—weed situation in the field, then the relative costs of the different products will probably be the deciding factor.

Crop safety

The safe use of herbicides is governed by crop species, variety, stage of growth, health and vigour of the crop and correct dose and application of the herbicide. Most herbicides can be applied to wheat and barley, relatively few to oats, rye, triticale and undersown crops. Many are restricted to certain varieties of cereal, and nearly all herbicides have to be applied within well defined crop growth stages. Even within the correct growth stages unusual weather and soil conditions can lead to herbicide damage. Early herbicide damage to cereals tends to be far less harmful to final yield than late herbicide damage. The latest safe stage for spraying the majority of herbicides is up to but before the one node detectable stage. However, there are opportunities for spraying weeds later than this as indicated in the relevant sections and tables that follow. For a full discussion of critical growth stages in cereals see Chap. 8 of The British Crop Protection Council *Weed Control Handbook: Principles*, 1982 Ed. H. A. Roberts, Blackwell, Oxford. Guides to growth stages are given in the Farmers Weekly Supplements and ADAS Booklets listed at the start of this chapter.

Between-Crops and Sowing Period

Annual grass weeds

Cultivations

The depth and type of primary cultivation has a strong influence on populations of annual grass weeds. Shallow cultivations and direct drilling encourage seedling emergence from seeds retained in the surface soil resulting in a rapid population build-up. Such a system is highly dependent on effective herbicides for control. With complete prevention of seeding shallow cultivations will result in the most rapid depletion of seed reserves.

Ploughing, by burying seeds deeply, encourages their persistence but results in fewer seedlings and is advised for systems of weed containment with low reliance on herbicides.

Early sowing in September will give less time for effective stubble cleaning and can result in more weeds and volunteer cereals in the next crop. These are likely to become well established in the autumn and require an autumn herbicide to avoid early competition with the crop. Autumn herbicides will be less necessary in late sown crops.

Where it is permissible and possible, thorough straw burning (straw swaths and intervening stubble) as soon as possible after harvest will give the maximum destruction to shed weed seeds. After burning cultivate shallowly to promote the germination of weed seeds which have lost dormancy, and to incorporate ash residues.

Stubbles should be cultivated shallowly as soon as possible after harvest to encourage the germination of weed seeds (unless perennial grass weeds are a more serious problem and require the use of post-harvest glyphosate). With autumn-sown cereals there is little opportunity to encourage natural seed losses by leaving stubble uncultivated. It is important that grass weed seedlings are not carried over as transplants into the next crop; they should be destroyed by re-cultivating (if dry) or by spraying with contact herbicide.

When it is intended to establish a cereal crop after a grass sward, spray glyphosate to kill all grass before ploughing. Plough to ensure good inversion and complete burial of the sward. This will prevent the entry of grasses such as rough meadow-grass, rye-grass and common couch which are often problems in the first cereal after grass.

Chemicals

Tri-allate may be applied in the liquid form as a soil-incorporated treatment in the 21 days before drilling wheat and barley. Tri-allate can also form the start of a useful sequence to control an extended range of weeds by applying one of the herbicides shown in Table 6.1 after drilling.

There are opportunities before drilling for using paraquat or glyphosate to control transplanted grass weeds surviving cultivations.

Annual broad-leaved weeds

Cultivations

The comments on cultivations and straw burning given above apply also to annual broad-leaved weeds except that seeds of broad-leaved species are generally more dormant than grass weeds so that shallow cultivations without ploughing are unlikely to result in rapid build-up of populations. Direct drilling systems will tend to exhaust the weed seeds in the surface soil, but enhance the survival of deeply buried seeds. Any change in the system to deep cultivation or ploughing will be more likely to re-establish a population of broad-leaved species than the less dormant grass weeds.

Chemicals

As for annual grass weeds there are opportunities before drilling or crop emergence of using paraquat with or without diquat, or glyphosate, to control weeds surviving cultivations.

Perennial grass weeds

Cultivations

An established method of control is to cultivate as soon as possible after harvest to fragment rhizomes so as to encourage the growth of dormant buds and exhaust the plant's food reserves. Shoot regrowth of 5–7 cm should be killed by further cultivation or sprayed with a contact herbicide such as paraquat, and the operation repeated if possible. Late harvesting of some crops together with the trend to early sowing of successive autumn cereal crops does not allow sufficient time for this strategy to be fully effective.

Chemicals

Chemicals available for use in this period are aminotriazole, glyphosate and TCA. Dalapon can be used but residual activity in the soil precludes the sowing of winter cereals after autumn applications.

Aminotriazole. This is a foliage-applied acting chemical which should be applied to actively growing couch foliage at least 7.5–10 cm in height. Stubbles should be left uncultivated to encourage re-growth. Plough 3–6 weeks after spraying, after which winter wheat or oats may be drilled. Aminotriazole is also recommended at a reduced rate without cultivation for low density couch where wheat is to be direct drilled.

Dalapon. This is also a foliage-applied acting chemical and the same recommendations apply as for aminotriazole except that dalapon residues are more persistent and winter cereals should not be sown. Delay sowing spring cereals until at least 40 days after ploughing.

Glyphosate. A translocated foliage-acting chemical which leaves no harmful residue in the soil. Use pre-harvest in the previous cereal, oilseed rape or pea crop, or leave stubbles undisturbed and delay application until couch re-growth has reached 10–15 cm. Cereals may be sown from 7 days after spraying.

TCA. This is a soil-acting chemical absorbed by plant roots and thorough incorporation with the soil is necessary after spraying. Leave 12–16 weeks before sowing any cereal.

Perennial broad-leaved weeds

Cultivations

Cultivations alone are seldom effective due to regrowth from fragmented underground parts of the plants. Reliance has therefore to be placed on chemical control. Shoots emerging from fragmented underground parts can be suppressed or controlled by using the herbicides mentioned in Table 6.4.

Chemicals

Use glyphosate pre-harvest or on leafy growth in disturbed stubble.

Control in a cereal/ley rotation

In a cereal/ley rotation it is important to control perennial broad-leaved weeds in grass before ploughing for wheat. Deep rooting species such as docks (*Rumex* spp.) will fragment and regenerate to infest the following cereal crop, and should be effectively killed with glyphosate when spraying the grass sward before ploughing, or with chemicals such as dicamba mixtures, asulam or triclopyr + clopyralid during the preceding summer.

Seed purity

It is important to sow cereal seed free from contamination with weed seeds. The greatest risk is with home-saved seed obtained from weedy fields where wild-oats (*Avena* spp.), brome grasses (*Bromus* spp.) or cleavers (*Galium aparine*) were harvested with the cereal crop. Such seed must be cleaned before sowing. Cereal seed samples can be tested for weed seed impurities at the Official Seed Testing Stations at Cambridge, Edinburgh or Belfast or at their satellite seed testing stations in the main cereal growing areas.

Straw disposal

Straw burning will destroy a proportion of seeds and reduce the dormancy of others, and so help to reduce the carry-over of weed seeds. Burn as soon as possible after harvest, having spread the straw to ensure a complete burn, and cultivate to incorporate ash and to encourage weed seed germination. If burning has to be limited, give priority to the most weedy fields.

Obtain baled straw to be used for animal feed or litter from clean fields to avoid the risk of spreading weed seeds via farmyard manure.

Systems of incorporating straw with tined implements alone are likely to increase weeds and will require a greater reliance on autumn herbicides. Ploughing offers good straw disposal and weed control providing the depth is not increased to return long buried but still viable seed to the surface.

Rotational ploughing

Continual shallow cultivations will increase annual weeds, especially grass weeds with relatively short dormancy. The surface soil will contain a high concentration of weed seeds, and straw and ash residues. Ploughing a proportion of the cereal area each year, say rotating the plough once every five years, can reduce weed emergence and improve herbicide efficiency.

Pre-emergence herbicides

A wide range of materials is available for the pre-emergence control of annual grass and broad-leaved weeds (Table 6.2). Pre-emergence control has the advantage of early removal of weed competition (see Chapter 2), and the drawback that the decision to spray at this time has to be based on past experience and the weed flora to be expected.

Crop Emergence to Tillering (Autumn/Winter) Period

This is an important period for the consideration of herbicide application to autumn-sown cereals, though low ground pressure vehicles or aircraft may be required to apply sprays at this time. High weed infestations, particularly of the early competing grass weeds, need to be sprayed in this period to obtain the maximum yield response. Early post-emergence treatments applied in the late autumn and winter are listed in Table 6.3. Choice of herbicide can be governed by the weed flora present. Many of the materials are both foliage- and soil-acting, with residual activity that extends control to spring germinating weeds.

Adverse soil and weather conditions, especially persistent frost may lead to crop tolerance problems. Assess crop vigour before undertaking winter herbicide spraying.

Tillering to Jointing (Spring) Period

Crops become sensitive to many broad-leaved weed herbicides at the beginning of jointing. In winter barley, spikelet initials develop during the winter, and jointing may commence in early March; wheat spikelets develop in the spring with jointing occurring in late April. This restricts the time available for safe and effective spring application of herbicides particularly for winter barley. Because of the dangers of late spraying it is preferable that broad-leaved weeds are controlled in winter barley in the autumn or winter; however, if this is not done and if bad weather delays spring spraying, select from the large number available those herbicides that are safe to spray after jointing commences (see Table 6.4).

TABLE 6.2 Herbicides for pre-emergence use in winter cereals

	Weeds controlled			*Use in winter crops of:*			
Active ingredient	Wild-oats	Black-grass	Broad-leaved weeds	wheat	barley	oats	rye
bifenox + linuron	No	No	Yes	√	No	No	No
chlorsulfuron + methabenzthiazuron	No	Yes	Yes	√	No	√	No
chlorsulfuron + metsulfuron-methyl	No	Some	Yes	√	No	√	No
chlortoluron	Yes	Yes	Some	√	√	No	No
chlortoluron + bifenox	Yes	Yes	Yes	√	√	No	No
isoproturon	Some	Yes	Some	√	√	No	√
isoproturon + triallate	Yes	Yes	Some	√	√	No	No
isoproturon + trifluralin	Some	Yes	Some	√	√	No	No
isoxaben	No	No	Yes	√	√	√	√
linuron	No	No	Yes	√	√	No	No
methabenzthiazuron	No	Yes	Yes	√	√	√	√
pendimethalin	No	Yes	Yes	√	√	No	√
pendimethalin + isoproturon	Some	Yes	Yes	√	√	No	√
terbutryne	No	Yes	Yes	√	√	√	√
tri-allate	Yes	Some	No	√	√	No	√
trifluralin ± linuron	No	No	Yes	√	√	No	No
trifluralin + terbutryne	No	No	Yes	√	√	No	No
trifluralin + trietazine + linuron	No	No	Yes	√	√	No	No

TABLE 6.3 Herbicides for post-emergence autumn/winter use in winter cereals

	Weeds controlled			*Use in winter crops of:*			
Active ingredient	Wild-oats	Black-grass	Broad-leaved weeds	wheat	barley	oats	rye
bentazone + dichlorprop + cyanazine	No	No	Yes	√	√	No	No
bifenox + mecoprop	No	No	Yes	√	√	√	No
bromoxynil + clopyralid	No	No	Yes	√	√	√	No
bromoxynil + clopyralid + mecoprop	No	No	Yes	√	√	√	No
bromoxynil + dichlorprop	No	No	Yes	√	√	√	No
bromoxynil + ioxynil	No	No	Yes	√	√	No	√
bromoxynil + ioxynil + benazolin	No	No	Yes	√	√	No	No
bromoxynil + ioxynil + isoproturon	Some	Some	Yes	√	√	No	No
bromoxynil + ioxynil + mecoprop	No	No	Yes	√	√	No	No
bromoxynil + ioxynil + trifluralin	No	No	Yes	√	√	√	No
chlorsulfuron + methabenzthiazuron	No	Yes	Yes	√	No	No	No
chlorsulfuron + metsulfuron-methyl	No	Some	Yes	√	√*	√	No
chlortoluron	Yes	Yes	Some	√	√	No	No
chlortoluron + bifenox	Yes	Yes	Yes	√*	No	No	No
clopyralid	No	No	Yes	√	√	√	No
clopyralid + bromoxynil + ioxynil + mecoprop	No	No	Yes	√	√	√	No
cyanazine + clopyralid	No	No	Yes	√	√	√	No
cyanazine + mecoprop	No	No	Yes	√	√	√	No

TABLE 6.3 Continued

	Weeds controlled			*Use in winter crops of:*			
Active ingredient	Wild-oats	Black-grass	Broad-leaved weeds	wheat	barley	oats	rye
diclofop-methyl	Yes	Yes	No	√	√	No	√
difenzoquat	Yes	No	No	√	√	No	√
fluroxypyr	No	No	Yes	√	√	No	No
fluroxypyr ± bromoxynil	No	No	Yes	√	√	No	No
ioxynil	No	No	Yes	√	√	No	√
ioxynil + benazolin + mecoprop	No	No	Yes	√	√	√	No
ioxynil + clopyralid	No	No	Yes	√	√	√	No
ioxynil + mecoprop	No	No	Yes	√	√	No	No
isoproturon	Some	Yes	Some	√	√	No	√
isoproturon + dichlorprop + bentazone	No	Yes	Yes	√	√	No	No
isoproturon + ioxynil + mecoprop	Some	Yes	Yes	√	√	No	No
isoproturon + ioxynil + bromoxynil + mecoprop	Some	Yes	Yes	√	√	No	No
isoproturon + trifluralin	Some	Yes	Some	√	√	No	No
methabenzthiazuron	No	Some	Some	√	No	No	No
mecoprop	No	No	Yes	√	√	√	No
mecoprop + clopyralid	No	No	Yes	√	√	√	No
metoxuron	No	Yes	Some	√	√	No	No
metoxuron + simazine	No	Yes	Some	√	√	No	No
metsulfuron-methyl	No	No	Yes	√	Some	No	No
tri-allate	Yes	Yes	No	√	√	No	No

Note:

Major product mixes only are listed. Many other herbicide mixtures are recommended on product labels.

*after 3 leaf stage only

TABLE 6.4 Herbicides for spring use in winter cereals for broad-leaved weed control

Active ingredients	*Use in winter crops of:*				*Recommended for use after jointing (Number of nodes present)*
	wheat	barley	oats	rye	
benazolin + 2,4-DB + MCPA	✓	✓	✓	No	No
bentazone + dichlorprop	✓	✓	✓	✓	Yes up to 2 nodes
bentazone + dichlorprop + cyanazine	✓	✓	No	No	No
bifenox + mecoprop	✓	✓	✓	No	Yes up to 1 node exceptionally 2 nodes
bromoxynil + ioxynil	✓	✓	✓	✓	Yes up to 2 nodes
bromoxynil + ioxynil + benazolin	✓	✓	✓	No	Yes up to 2 nodes
bromoxynil + ioxynil + isoproturon	✓	✓	No	No	Yes up to 1 node
bromoxynil + ioxynil + mecoprop	✓	✓	✓	No	Yes up to 1 node
bromoxynil + ioxynil + trifluralin	✓	✓	✓	No	Yes up to 2 nodes
clopyralid	✓	✓	✓	No	Yes up to 2 nodes
clopyralid + bromoxynil ± MCPA or mecoprop	✓	✓	✓	No	Yes up to 1 node
clopyralid + bromoxynil + mecoprop	✓	✓	✓	No	Yes up to 1 node exceptionally 2 nodes
clopyralid + bromoxynil + ioxynil + mecoprop	✓	✓	✓	No	Yes up to 2 nodes
clopyralid + ioxynil	✓	✓	✓	No	Yes up to 2 nodes
clopyralid + ioxynil + mecoprop	✓	✓	✓	No	Yes up to 1 node exceptionally 2 nodes
chlorsulfuron + bromoxynil + ioxynil	✓	✓	No	No	Yes up to 2 nodes
cyanazine + MCPA	✓	✓	✓	No	No
cyanazine + mecoprop	✓	✓	✓	No	Yes up to 1 node exceptionally 2 nodes
2,4-D	✓	✓	✓	✓	No
2,4-DB	✓	✓	✓	✓	No
2,4-DB + linuron + MCPA	✓	✓	✓	No	No

TABLE 6.4—Continued

	Use in winter crops of:				
Active ingredients	wheat	barley	oats	rye	*Recommended for use after jointing*
2,4-DB + MCPA	√	√	√	√	No
dicamba + dichlorprop + MCPA	√	√	√	No	No
dicamba + mecoprop	√	√	√	√	No
dicamba + mecoprop + MCPA	√	√	√	No	No
dichlorprop	√	√	√	No	No
dichlorprop + 2,4-D	√	√	√	No	No
dichlorprop + MCPA	√	√	√	No	No
dichlorprop + MCPA + clopyralid	√	√	√	No	No
dichlorprop + MCPA + ioxynil	√	√	√	No	No
dichlorprop + MCPA + 2,3,6-TBA	√	√	√	No	No
dichlorprop + mecoprop	√	√	√	No	No
dinoseb amine, acetate	√	√	√	No	No
fluroxypyr	√	√	√	No	Yes up to flag leaf emergence
fluroxypyr + ioxynil	√	√	No	No	Yes up to 2 nodes
fluroxypyr + ioxynil + bromoxynil	√	√	No	No	Yes up to 1 node exceptionally 2 nodes
isoproturon + dichlorprop + bentazone	√	√	No	No	Yes up to 1 node
isoproturon + ioxynil + mecoprop	√	√	No	No	Yes up to 1 node
MCPA	√	√	√	√	No
MCPB	√	√	√	√	No
MCPB + MCPA	√	√	√	√	No
mecoprop	√	√	√	No	Yes up to 1 node exceptionally 2 nodes
mecoprop + clopyralid	√	√	√	No	Yes up to 1 node exceptionally 2 nodes
mecoprop + ioxynil + benozolin	√	√	√	No	Yes up to 2 nodes
mecoprop + MCPA + 2,3,6-TBA + dicamba	√	√	√	√	No
metsulfuron-methyl	√	Some	No	No	Yes up to 2 nodes

Note: Major product mixes only are listed. Many other herbicide mixtures are recommended on product labels.

Control of late germinating or surviving wild-oats and blackgrass is possible at this time (see Tables 6.5 and 6.6). Early in this period diclofop-methyl or difenzoquat can be used for wild-oat control, and chlortoluron, isoproturon or metoxuron if black-grass predominates. Later in spring as winter wheat or winter barley reach the leaf sheath erect stage (GS 30) the choice of material extends to include benzoylprop-ethyl for wild-oats and flamprop-M-isopropyl or flamprop-methyl for wild-oats together with some control or suppression of black-grass. Diclofop-methyl also gives some control of seedling black-grass although primarily used for wild-oat control.

Seedlings and young plants of several other grasses are controlled by a number of these herbicides as detailed under the individual chemicals in Chapter 4.

TABLE 6.5 Herbicides for spring use in winter cereals for wild-oats and black-grass control

	Spring control of:		*Use in winter crops of:*			
Active ingredient	wild-oats	black-grass	wheat	barley	oats	rye
benzoylprop-ethyl	Yes	No	√	No	No	No
chlortoluron	No	Yes	√*	√*	No	No
diclofop-methyl†	Yes	Some	√	√	No	√
difenzoquat	Yes	No	√*	√	No	√
flamprop-M-isopropyl	Yes	Some	√	√	No	No
flamprop-methyl	Yes	Some	√	No	No	No
isoproturon	Some	Yes	√	√	No	No
isoproturon + ioxynil + bromoxynil	Some	Yes	√	√	No	No
isoproturon + ioxynil + mecoprop	Some	Yes	√	√	No	No
metoxuron	Some	Yes	√*	√*	No	No
metoxuron + simazine	Some	Yes	√*	√*	No	No

†excluding northern England and Scotland *with varietal restrictions

Period After Two Nodes Detectable.

By this time weed control should normally be complete but where prevention of seeding is important a late application of spray can be considered.

For late wild-oat control benzoylprop-ethyl and flamprop-M-isopropyl may be applied up to the fourth node detectable stage of the crop.

For late broad-leaved weed control cyanazine plus clopyralid with or without fluroxypyr can be applied up to the third node detectable stage. Clopyralid plus ioxynil mixed with fluroxypyr is recommeded for use after the second node but before the third node detectable stage; these mixtures have good activity on mature mayweed, corn marigold and thistles. Fluroxypyr improves the control of large chickweed and cleavers and can be used alone up to the time of flag-leaf emergence in winter wheat and winter barley.

TABLE 6.6 Application timing of herbicides used mainly for wild-oats and/or black-grass control in cereals

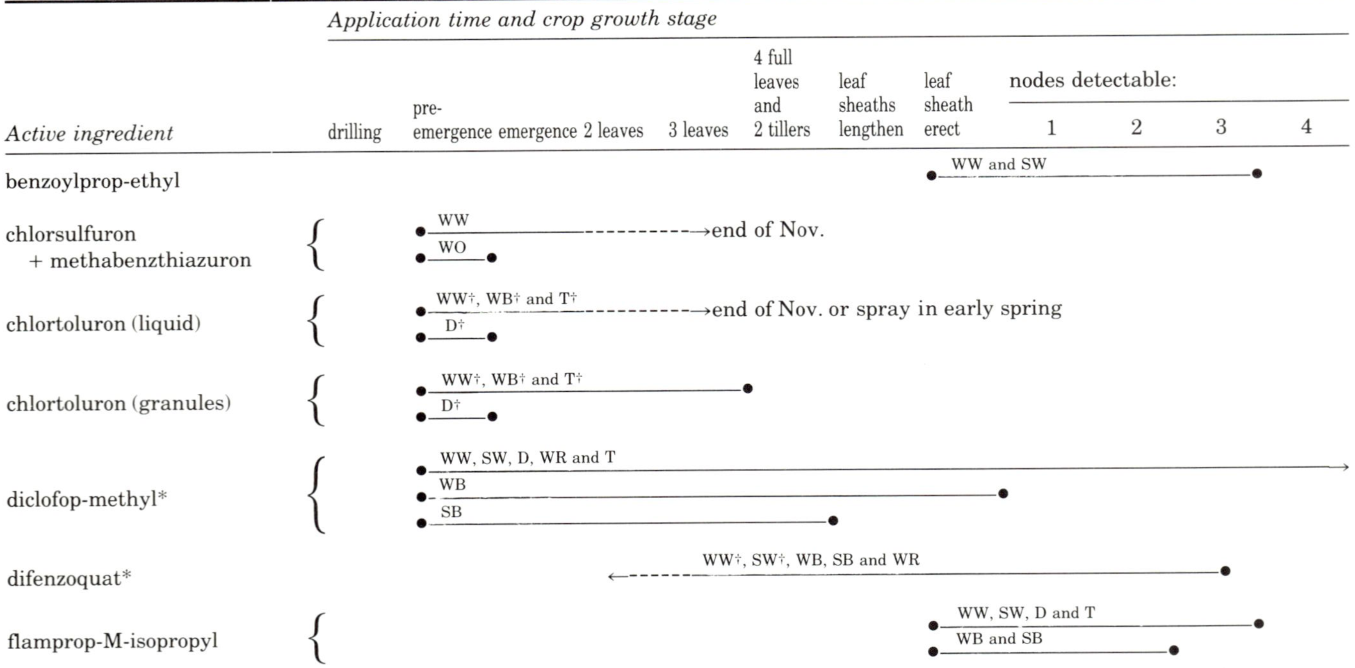

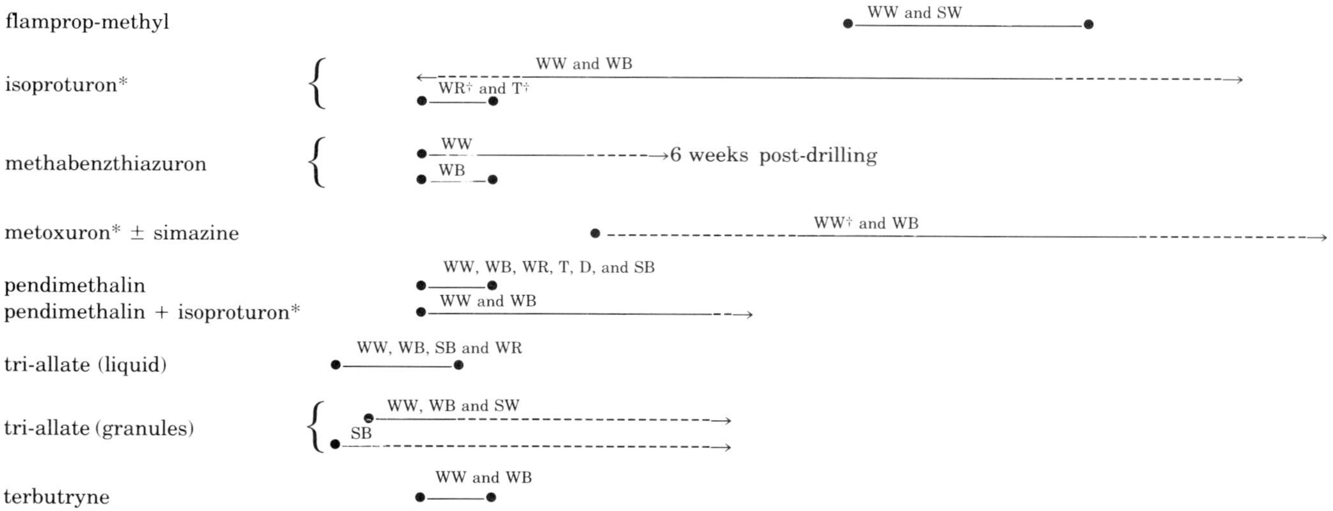

Notes: *application timing based on weed emergence and growth stage within the crop stage shown
†certain varieties only
WW: winter wheat
WB: winter barley
WO: winter oat
WR: winter rye
SW: spring wheat
SB: spring barley
D: durum wheat
T: triticale

Recommendations for Control of Specific Grass Weeds

Wild-oats

The ability to hand-rogue wild-oats (*Avena fatua* and *Avena lodoviciana*) together with the availability of efficient herbicides provides a choice between containing the weed at minimum cost and pursuing a policy of eradication where hand-roguing supplements the use of herbicides. The effective use of herbicides should reduce high infestations to roguable levels in three or four years; eradication of low levels is more difficult and roguing needs to be thoroughly carried out. With cereal seed crops an eradication policy is usual; here roguing is essential because wild-oats are economically damaging at very low cost levels.

When a policy involving both hand roguing and chemical control is planned it must be borne in mind that all the wild-oat herbicides except tri-allate and chlortoluron stunt partly controlled survivors.

The protracted emergence of wild-oats complicates the optimum timing for herbicide application. Plants emerging with the crop will be most competitive and require an autumn or winter herbicide to avoid yield damaging competition in the early spring.Herbicides at this time will not control spring emerging wild-oats which although suppressed by the crop will still form seeds. Such seeding may be ignored if the policy is one of containment but is very significant if the aim is to eradicate the weed. In this case a late spring herbicide is needed, either in sequence to an autumn herbicide to control high populations where yields are at risk, or alone when wild-oats numbers are low, to minimise seed return.

More details of cultural and chemical control are to be found in ADAS Leaflet 452, *Wild Oats*.

Black-grass

Black-grass (*Alopecurus myosuroides*) is a prolific seed producer and is greatly encouraged by shallow cultivations and direct drilling. With these systems a high degree of control, better than that required for wild-oats, is needed to prevent populations from increasing. A sufficiently high and consistent level of control by herbicides is not likely, especially if herbicide performance is reduced by ash and trash residues. Rotational ploughing as described earlier in this chaper should, by burying seeds and residues, reduce the population and improve herbicide performance.

In the absence of ploughing an integrated control strategy is essential. Unlike wild-oats, black-grass cannot be hand rogued to supplement herbicide control. Husbandry practices which can improve the overall control are:

a) efficient straw burning to destroy weed seeds;

b) drilling later rather than early in the autumn to give time to destroy the main flush of weed seedlings;

c) stubble hygiene to achieve complete weed kill before drilling;
d) growing competitive crops, by using adequate amounts of seed and fertiliser and ensuring that the land is well drained;
e) including a run of spring-sown broad-leaved crops in the rotation.

More details of cultural and chemical control are to be found in Leaflet 522, *Black-grass*.

Meadow-grasses

Meadow-grasses (*Poa annua* and *Poa trivialis*) are prolific seed producers and are associated with shallow cultivations and high moisture conditions; they are more prevalent in cereals in high rainfall areas and in cereals rotated with grass. Where good, clean inversion ploughing is not possible it is necessary to rely on herbicides for control (Tables 6.1 and 6.2). They are susceptible to those herbicides which control black-grass except that diclofop-methyl only controls rough meadow-grass.

Unless a stubble spray of glyphosate or paraquat has been used, beware of cultivations which leave transplants of rough meadow-grass near the surface. These can persist as mature plants which are hard to kill with herbicides that are selective in cereals.

Annual meadow-grass germinates throughout the year; rough meadow-grass peak germination in late spring and late autumn. Rotating winter cereal with spring barley allows stubbles to be left to encourage seed losses and prevent a build-up of meadow-grass.

Brome grasses

If possible burn straw to destroy brome seeds. Brome is mostly a hedge and field margin weed so headland firebreaks need special treatment. Chemical control of brome is often unreliable and needs integrating with cultural measures. Brome seeds are susceptible to burial, so plough deeply; complete soil inversion gives good control.

Sequences of tri-allate followed by metoxuron or isoproturon give the most reliable chemical control. Meadow brome (*Bromus commutatus*) emerges somewhat later in the autumn and is harder to kill with herbicides than barren brome (*Bromus sterilis*).

Brome grasses produce many seeds which are of short persistence in the soil. Brome has become a problem with early drilled winter cereals; changing the rotation to include spring crops or break crops such as oilseed rape or grass in which brome can be effectively controlled will help to break the annual seed production cycle and should dramatically reduce the population. More details of cultural and chemical control are to be found in ADAS Leaflet 777, *Barren Brome*.

Awned canary-grass

Awned canary-grass is similarly favoured by early winter cropping and by reduced cultivations. As with brome, infestations are usually confined to the headlands, but seeds can persist for longer in the soil. Herbicides which give good control are pendimethalin, terbutryne and diclofop-methyl; awned canary-grass is not well controlled by the urea herbicides. Further details are given in ADAS Leaflet 915, *Awned canary-grass.*

Couch grasses

Common couch (*Elymus repens*) and Black bent (*Agrostis gigantea*) are propagated to some extent by seeds but mainly by underground rhizomes. For effective long-term control it is necessary to kill the rhizome system, either by a system of chopping and exhausting the rhizomes, or by translocated herbicides. Translocation and rhizome kill is best achieved when foliage-applied herbicides are applied to the maximum area of actively growing foilage. Autumn treatments (see p. 115) are not always successful due to inadequate regrowth in the short time between harvesting one crop and sowing the next. Glyphosate applied to the mature crop before harvest, when the couch is growing actively, will give more reliable control and is advised for heavy infestations. Further details are given in ADAS Leaflet No. 89, *Couch.*

Onion couch (False oat-grass)

Onion couch (*Arrenatherum elatius* var *bulbosum*) is propagated partly by seeds but most importantly by the swollen internodes or "strings" of bulbs at the base of the stems, just below soil level. These bulbs are resistant to drought. Direct drilling or shallow cultivations will favour onion couch by separating and dispersing the individual bulbs. Efficient ploughing to bury the bulbs will give good control. For heavy infestations integrate ploughing with chemical control. The wild-oat herbicides, benzoylprop-ethyl, flamprop-methyl and flamprop-isopropyl will control aerial growth and give some kill of bulbs; pre-harvest glyphosate will give a better kill of bulbs if there is sufficient green leaf present at spraying. Autumn herbicides are less successful as re-growth after harvest is slower than with common couch.

Futher information on cultural and chemical control is to be found in ADAS Leaflet 89, *Couch.*

DISEASE CONTROL

The need for integrated methods

Recent events have underlined the fact that fungicides alone do not provide the complete answer to disease problems: there has been an upsurge in diseases such as take-all and barley yellow mosaic virus (BaYMV) for which there is as yet no effective chemical control. Futhermore fungicide resistance in some pathogens and definite, though less dramatic, loss of sensitivity in others, have highlighted the dangers of relying solely on fungicides for disease control. The integration of chemical, cultural and genetic methods is particularly important in controlling cereal diseases.

Fungicide resistance

The occurrence of mercury resistant strains of *Pyrenophora avenae* has meant that since the late 1960s it has not been possible to rely on organomercury seed treatments for the control of the relatively unimportant leaf spot disease of oats. In the 1970s the effectiveness of ethirimol declined as mildew strains (*Erysiphe graminis*) arose which were less than fully sensitive to this fungicide. Recently mildew strains showing reduced sensitivity to the so-called Group 2, dimethylation inhibitor, fungicides (see Table 4.7) have become increasingly common. Although, as with the ethirimol, there has been no general breakdown in disease control, these materials often now fail to give the very long-term protection against infection which they once provided. In addition strains of the eyespot fungus (*Pseudocercosporella herpotrichoides*) completely resistant to the Group 1, benzimidazole, fungicides are now widespread in the main cereal growing areas.
The risk of fungicide resistance must be borne in mind when planning a disease control strategy. The frequent application to a crop of the same fungicide, or of fungicides with the same mode of action, should be avoided. Except where absolutely necessary late season (post-flowering) application of fungicides should also be avoided as these exert selection pressure on that part of the pathogen population which will survive until the following season. Wherever possible full use should be made of multi-site fungicides (which are less prone to resistance problems) and of formulated mixtures, or label recommended tank mixes of fungicides, each component of which is active against the disease or diseases concerned.

Fungicide usage

Within an integrated control strategy the use of fungicides is likely to remain the mainstay of disease control. Three distinct approaches to fungicide use are possible:–

Disease risk assessment system. This system is based on the regular monitoring of each crop, fungicides being applied only when "threshold" levels of disease are found or when weather conditions are such that the spread of a certain disease (e.g. septoria) appears imminent. Use of this method will keep fungicide costs to a minimum, but it can underestimate the need to spray.

Routine prophylactic fungicide application. This method involves the routine application of fungicide at specific growth stages (e.g. late tillering, flag leaf emergence and/or ear emergence) without reference to disease levels in the crop. The method offers obvious management advantages to those farming a large acreage of cereals, but it is inflexible and leads to inappropriate timing, to over-use, or (occasionally) to under-use of fungicides.

Managed disease control system. This system attempts to combine the best features of the other two, encouraging crop monitoring and fungicide application in response to disease risk situations but planning for routine sprays to be applied at certain growth stages.

All three systems of fungicide use are describes in ADAS Booklet 2257, *Use of fungicides and insecticides on cereals,* MAFF Publications, Alnwick. The use of fungicides described in this chapter is based on the managed disease control systems which are presented in the form of decision trees in ADAS Leaflet 831, *Winter wheat managed disease control*; Leaflet 843, *Winter barley managed disease control*; and Leaflet 844, *Spring barley managed disease control*, MAFF Publications, Alnwick. The information is also presented in the Farmers Weekly *Cereal Fungicide* supplement published in association with ADAS.

Fungicide programmes

If, for management reasons, it is decided to apply a routine programme to high potential crops in intensive cereal situations the best returns are likely to be achieved from an eyespot spray at early jointing followed by a broad-spectrum fungicide at the flag leaf to ear emergence stage. Additional sprays may be added to this basic programme if the development of particular diseases demands them, but the *routine* application of three or more sprays is unlikely to be cost effective.

Herbicide and growth regulator/disease interactions

A few herbicides have direct effects on plant pathogens; difenzoquat for example, will give a reasonable control of mildew (*Erysiphe graminis*) and glyphosate will suppress the sporulation of the net blotch fungus (*Pyrenophora teres*) on leaf debris. Indirect effects of herbicide usage may be beneficial; the control of volunteers and weed hosts can reduce the carry-over of obligate parasites. On the other hand mis-application of some herbicides can weaken the plants making them more susceptible to take-all attack and other diseases.

Pre-harvest application of glyphosate to control couch grass can if carried out before a break crop reduce the carry over of take-all. In the short term, however, the take-all fungus (*Gaeumannomyces graminis*) builds up on the dead rhizomes so if the chemical is applied in a continuous cereal situation the subsequent crop can suffer particularly severely from the disease.

The use of growth regulators which reduce internode length increases the susceptibility of plants to splash borne diseases such as septoria (*Septoria nodorum* and *S. tritici*) and rhynchosporium (*Rhynchosporium secalis*). On the other hand, while these materials have no direct effect on eyespot incidence, they may by strengthening the straw reduce the lodging caused by this disease.

Between-Crops and Sowing Period

Cultural methods

Crop sequence

Cropping sequence has a critical effect on a number of diseases. Take-all can be well controlled by a one year break from susceptible cereals or, alternatively, its worst effects can be overcome by the exploitation of take-all decline in continuous wheat and barley sequences as described in the individual disease section towards the end of this chapter. The control of eyespot normally requires a longer break from wheat or barley as debris ploughed under before the break-crop is still infective when ploughed up again after it. Sharp eyespot (*Rhizoctonia cerealis*) is more favoured by ley/arable systems than by continuous cereal growing. Grass leys ploughed-in late can also increase the risks of barley yellow dwarf virus (BYDV) attack in subseqeunt cereal crops, and wireworm damage after ploughed-out grass can increase the incidence of cephalosporium leaf stripe.

Stubble hygiene

Obligate parasites, such as the rusts, which need green plant tissue on which to survive, are favoured by continuous cereal cropping only if infected volunteer plants survive to carry the pathogens from one crop to the next. It should be remembered, however, that even if ploughing or herbicide application destroys these volunteers before the next crop is sown they may already have acted as a source of disease for earlier sown crops on adjacent fields. The need for good field hygiene to remove the possibility of a 'green bridge' carrying leaf diseases over from one crop to another is paramount.

Straw disposal

Diseases as such as septoria leaf spot and net blotch which can survive on dead plant debris present a different kind of hygiene problem.

Application of herbicides to the stubble has less effect on these – though glyphosate does appear to suppress the sporulation of the net blotch fungus to some degree. Straw burning or, even more effectively, incorporation by ploughing, will reduce but not eliminate fungi at the soil surface. At best, therefore, the application of these techniques will only delay the development of these diseases in the subsequent cereal crops, but this delay will be of value if it removes the need to spray against a disease such as net blotch in the autumn.

Eyespot on the stem bases and, more particularly, take-all on the roots are not directly affected by straw burning. Take-all may, however, be aggravated by straw incorporation if this produces a loose seedbed which favours the spread of the fungus over the roots.

Variety selection

There is a wide range of variation in varietal susceptibility to most of the major cereal diseases. Avoid varieties that have a low level of resistance to diseases which past experience has shown to present problems on the farm. The selection of varieties can often be linked to particular field situations. It would be unwise, for example, to grow a mildew susceptible variety on the highly organic soil of recently ploughed out grassland, or a septoria susceptible one in a valley bottom where summer mist hangs.

From time to time pathogen strains arise which are well adapted to varieties which were previously highly resistant to particular diseases. This is more likely to occur where a single gene has conferred a very high level of resistance on the plants than where a number of genes work together to produce a moderate level of "field" resistance. For this reason it can be counter-productive to rely exclusively on one particular variety however highly it is rated for resistance to the major pathogens. It is better to grow a number of varieties with different genetic base for their disease resistance. So far as mildew and yellow rust (*Puccinia striiformis*) of winter wheat are concerned varieties can be assigned to "diversification groups", each group being resistant to the strains of the pathogen which occur on the others. It thus becomes possible to grow, for example, three winter wheat varieties selected from three different diversification groups with respect to, say, yellow rust. Only if three rust races (each virulent on a different one of the three diversification groups) occur in the area will all the crops on the farm succumb to the disease. A table showing the diversification grouping is given in MAFF Booklet 2257, *Use of fungicides and insecticides in cereals*, and the Farmers Weekly *Cereal Fungicides* supplement published in association with ADAS. If instead of being grown separately the three varieties are grown as a blend, the overall effects of the disease can be even further reduced. During the rapid phase of disease development most new lesions develop mainly as a result of infection by spores produced within the crop itself. In a three-way blend the spores will have only about a 1 in

3 chance of landing on the leaf of a variety which it will be able to attack. Thus, even if all three of the appropriate rust races are present in the area disease levels in a blend of the three selected varieties will be significantly less than the average of levels in the three varieties grown separately. However, the demands of the market for the cereals must be borne in mind. Blended parcels of differing varieties frequently suffer reduced sales potential.

Seedbed conditions

Avoid deep 'puffy' seed beds. The effects of a loose seedbed on take-all have already been mentioned in the context of straw disposal and the need, especially on light and organic soils, to achieve a good firm seedbed cannot be too strongly emphasised.

Sowing too deeply weakens plants and aggravates the effects of an early take-all attack, and of pathogens such as *Fusarium* spp. and *Rhizoctonia cerealis* which can rot the sub-crown internode.

Shallow drilling produces a plant which, during the tillering stages, is more prostrate in habit and in which the length of the white underground portion of the tiller base is reduced to a minimum. This combination of characteristics appears to render the plants far less vulnerable to attack by the eyespot fungus and may explain why direct drilled crops often suffer less severely from eyespot than their generally more deeply drilled counterparts sown after ploughing.

Time of sowing

Time of sowing has a profound influence on disease incidence. Early sown crops emerge at a time when temperatures are still high enough to favour infection by both root and foliar pathogens, when unploughed stubbles with infected volunteers or unharvested cereals will still be present in the neighbourhood to act as sources of inoculum, and when viruliferous aphids will still be on the wing. The risks of autumn infection by both fungi and aphid borne viruses are increased by early sowing, though with diseases such as eyespot and BYDV the symptoms may not show up until the following spring.

Certain complicating factors must, however, be borne in mind when considering the effects of sowing date on disease. Firstly, early-sown winter barley crops which suffer particularly severe attacks of mildew in autumn often grow away from the disease most rapidly in spring, so summer disease levels can be higher on later sown crops which suffered only light attacks during the autumn and winter. Secondly, and again especially in the context of barley mildew, early-sown crops are often so well tillered that the control of autumn mildew, and the subsequent increase in crop vigour and tiller survival, can sometimes produce too thick a crop. Too high a population of fertile tillers can, especially in years of summer drought, result in a reduction in grain size and occasionally in marked decrease in harvested yield.

Crop nutrition

It is most important to maintain a well balanced regime of fertiliser application. Crops deficient in potash or, even more particularly, phosphate are, for example, especially vulnerable to attack by take-all, and manganese deficient plants are particularly susceptible to mildew. Adequate spring nitrogen helps plants to cope with root loss caused by take-all; but excessive nitrogen renders them more susceptible to most foliar diseases.

Take-all may be aggravated by over-liming, which favours the fungus, or by pH levels which, though not so low as to inhibit fungal growth, are low enough to weaken the roots and render them more susceptible to attack. Avoid the development of acid patches by regular liming where necessary, planning to avoid applying lime immediately before the second to fourth cereals after a break when the risk of severe take-all is highest.

Treatment against seed-borne disease

Of the 16 or so seed-borne diseases of cereals in the U.K. only two, the loose smuts of wheat and barley (caused by strains of *Ustilago nuda*) arise from pathogens carried deep in the seed. All the others are found superficially in or on the seed coat.

Control of the pathogens carried externally on the seed can be achieved, with varying degrees of effectiveness, with an inexpensive organomercury seed treatment. The wheat and barley loose smuts are not controlled by this material and for their effective control in valuable seed stocks more expensive systemic products are needed. These may be smut specific materials used in conjunction with organomercury, or broad-spectrum mercury replacements such as those shown in Table 6.7. Loose smut of winter barley is well controlled by triadimenol plus fuberidazol plus imazalil. Loose smut in wheat is best controlled with carboxin plus organomercury or triadimenol plus fuberidazole

In wheat, bunt or covered smut (*Tilletia caries*) is extremely well controlled by organomercury and all the other chemicals listed for use on wheat in Table 6.7 so nowadays it only rarely occurs, usually where a seed stock has been left untreated for a number of seasons. Similarly, covered smut of barley(*Ustilago hordei*) is controlled by all the seed treatments so is only occasionally found.

Foot rots and seedling blights of wheat and barley, caused by seed-borne *Cochliobolus sativus*, *Septoria nodorum* (on wheat) and *Fusarium* spp. sometime affect crop establishment. The latter two are only partially controlled by organomercury; they are well controlled on wheat by carboxin plus thiabendazole and by tridimenol plus fuberidazole. All treatments give partial control.

Barley is specifically attacked by two seed-borne leaf diseases, net blotch (*Pyrenophora teres*) and leaf stripe (*P. graminea*). Most infection of net blotch is due to spores produced on debris, but seed-borne inoculum is important in introducing the disease into otherwise clean areas; all seed treatments give only partial control. Leaf stripe, which can lead to yield losses by killing seedlings or reducing plant vigour and causing ear blindess, is rarely seen, being well controlled by all the seed-treatments recommended for winter barley except tridimenol plus fuberidazol. Before the introduction of organomercury, leaf stripe was a serious disease and could become so again if effective seed treatments were not used. There is evidence that *Rhynchosporium secalis* (the cause of leaf blotch) can be carried on the seed and this is only partially controlled by seed treatments.

Leaf spot of oats (*Pyrenophora avenae*) is an example of a disease where mercury-resistant strains of the pathogen are widespread. Despite this fact, leaf spot rarely causes serious yield losses. The covered and loose smuts of oats (*Ustilago hordei* and *U. avenae*) are both well controlled by organomercury seed treatment, as is the stripe smut of rye (*Urocystis occulta*); the latter disease is rare and in practice organomercury seed treatments are seldom applied to rye.

Some of the mercury substitutes now available for use as seed treatments have a similar spectrum of activity to organomercury itself, others control the loose smuts as well. Others offer additional benefits in that they are taken up by the roots and afford protection against leaf diseases as indicated in Table 6.7.

Potential seed crops should have been inspected for all seed-borne diseases during the growing season, especially after ear emergence when the smuts will be visible. Only save seed from a crop containing more than 0.5% loose smut if it can be treated with a fungicide which will control the disease.

All bought-in seed for grain production should be treated at least with one of the less expensive seed treatments. Home-saved seed, even from apparently clean crops, should be treated at least every other year and preferably every year. Application requires purpose built machinery. All seed from the farm should be treated in any one year.

Seed treatments for foliar disease

Prophylactic treatments of fungicides applied to seed may be useful to control barley mildew in high risk situations such as susceptible varieties sown in September on light land (Table 6.8). A number of other diseases are controlled by these treatments as indicated in the table.

When a seed treatment has been used against mildew and the disease develops on the growing crop later, use a spray with a different mode of action (see p 86).

TABLE 6.7 Fungicides for the treatment of seed-borne diseases

	Use on seed of:–				
Active ingredients	wheat	barley	oats	rye	triticale
organomercury	√	√	√	√	√
carboxin + organomercury	√	√[1]	√		
carboxin + thiabendazole	√	No	No	√	
carboxin + thiabendazole + imazalil	No	√[1]	√		
flutriafol + ethirimol + thiabendazole	No	√[4]	No		
guazatine	√	No	No		
guazatine + imazalil	No	√	√		
triadimenol + fuberidazole	√[3]	√[5,6]	√	√	
triadimenol + fuberidazole + imazalil	No	√[2,5]	No		

Notes
[1]spring barley only
[2]winter barley only
[3]some control of mildew, septoria and snow rot also
[4]some control of mildew, rhynchosporium and rusts also
[5]as [4] plus some control of snow rot also
[6]no control of leaf stripe in winter barley

Crop Emergence to Tillering (Winter) Period

Barley: application of fungicides in the autumn

In early-sown fields where barley was grown in the previous season, and especially where much straw debris has been left on the surface, net blotch may occasionally be present at sufficiently high levels to require treatment. Apply an appropriate fungicide if, on average, the disease covers 10% of one leaf per plant. In general, however, mildew is the disease most likely to require attention in the autumn. Yield responses with autumn spraying will vary with the severity of the subsequent weather, being greater after mild, open winters. They are also influenced by the following factors:

Sowing date and crop density. Mildew levels are generally highest on early sowings. On the other hand, if mild weather in early winter allows late-sown crops to become infected they may suffer particularly severely from the effects of the disease. Crops sown in early September may be so thick that application of autumn fungicide will occasionally reduce yield by allowing too many tillers to survive, with consequent decrease in grain size.

TABLE 6.8 Seed treatments for barley mildew control

Active ingredient
ethirimol
flutriafol + ethirimol + thiabendazole[1]
triadimenol + fuberidazole[1,2]
triadimenol + fuberidazole[1,2] + imazalil[1,2] (winter barley only)
triforine

[1]also controls rhynchosporium with some control of rusts
[2]also partially controls rhynchosporium, brown rust and yellow rust
All these treatments except triforine also control important seed-borne diseases. (See Table 6.7.)

The variety. The more mildew susceptible the variety the greater the likely response. It should be noted, however, that even fairly resistant varieties can suffer quite severe attacks if conditions are favourable to early development. Mildew increases frost susceptibility so it is particularly important to control it on spring varieties sown in the autumn and on frost susceptible winter varieties such as Maris Otter.

Soil type. Crops on light land suffer more severely from frost damage (which mildew will accentuate), and the poorer rooting resulting from autumn mildew attacks makes such crops more susceptible to drought stress in spring and summer. In addition it should be noted that mildew levels in early autumn are usually higher on light land crops than on otherwise comparable crops on heavy soils

The application of fungicides to barley in the autumn may be rationalised as follows:–

- Thin crops on medium soils and all except the thickest crops on light land should be sprayed as soon as active mildew pustules are noticed on the lower leaves.
- Thick crops on medium soils and crops on heavy land should be sprayed if mildew becomes obvious on the youngest leaves.
- Spring varieties sown in the autumn should be sprayed if disease (mildew, net blotch and/or rhynchosporium) is present throughout the crop.
- A broad-spectrum fungicide applied for mildew control in late autumn sometimes retards the development of rhynchosporium and other diseases later in the winter.
- The prophylactic use of a mildew fungicidal seed treatment should be considered as an alternative to autumn spraying in situations where past experience suggests the disease levels are likely to be serious. (see "Seed treatment for foliar disease" section above).

Wheat: eyespot and mildew-control in winter

Eyespot. Inspect early sown crops of susceptible varieties for eyespot in late January and February especially after mild winters. Ignore late, thin tillers which will obviously not produce an ear. Strip off the sheaths of dead leaves from the remaining tillers, and if on 30% of these tillers lesions are found on the sheaths of the still green leaves use an eyespot spray (see Table 6.9). A second spray may be needed later in the season to maintain control.

Mildew. If mildew is well established on the upper leaves at this time an appropriate fungicide should be applied. (Table 6.9).

Barley: fungicide application in late winter/early spring

If late winter examination of the crop shows the majority of plants to be carrying obvious disease on the upper leaves (say 5–10% of the top leaf covered with mildew and/or rhynchosporium) then a spray should be applied before spring growth begins.

Tillering to Jointing (Spring) Period

Wheat: leaf sheath erect to early jointing

Eyespot. Examine the crop at the leaf sheath erect to first node visible growth stage (usually late March to mid-April) and spray if 20% of the stems bear lesions (assess disease only on those tillers which are similar in growth stage to the main shoot). Early-sown crops in intensive cereal situations are most likely to require spraying.

If regular usage of benzimidozole (MBC) fungicides (benomyl, carbendazim or thiophanate-methyl) in previous years suggests that MBC-resistant strains of the eyespot fungus are likely to have been selected, then these materials should only be used in combination with a fungicide active against eyespot but having a different mode of action such as prochloraz or propicanazole.

Mildew. Materials applied to control eyespot will give some control of low levels of mildew. If active mildew pustules are obvious on the green (i.e. not senescent) leaves of most plants in the crop, which may occur at this time after a mild winter especially on highly organic soils or where manganese deficiency symptoms occur, apply a mildew specific fungicide together with the eyespot spray.

Septoria. Fungicides used to control eyespot are also active against *S. tritici*. On very susceptible varieties, and in coastal areas, it may sometimes be necessary, if obvious lesions are present on the lower green leaves, to spray against this disease even if no treatment is necessary against eyespot.

Yellow rust. Highly susceptible varieties (NIAB rating 3 or less) should be sprayed when the disease is first seen in the crop.

Barley: leaf sheath erect to early stem extension stage

Eyespot. Early-sown crops in intensive cereal situations are likely to be infected. Crops should be examined at the leaf sheath erect to first node visible stage. Diagnosis can be difficult due to high pigmentation of the lower leaf sheath, rhynchosporium lesions on the sheaths, and generalised browning caused by *Fusarium* spp. If on more than 20% of the shoots (ignoring those tillers which will obviously not produce ears) brown lesions are found to be penetrating beyond the outer leaf sheaths then an eyespot fungicide should be applied.

If the routine use of MBC fungicides in previous years suggests that MBC-resistant strains of fungus are likely to be present then these materials should be used only in combination with other fungicides active against eyespot. See Table 6.9.

Leaf diseases. Early-sown crops which suffer severe autumn mildew infection may lose so much leaf as a result of frost damage that they carry very little of the disease through the winter, and they often grow away very quickly from any disease that has survived. Late-sown crops may escape autumn infection but they tend to suffer more from mildew in spring. Rhynchosporium, net blotch and brown or yellow rusts may also be found on winter barley at the beginning of spring growth. While the obvious presence of any of these pathogens will determine which fungicide is used for application at this stage, even crops in which disease levels are at a very low level may be expected to respond to a broad-spectrum material applied at the leaf sheath erect to the first node visible stage. It is advisable to use fungicides with a different mode of action from any used in the autumn or late winter.

Oats: application of fungicides at late tillering stage

If mildew is present when a spring herbicide is to be applied then a compatible mildew fungicide should be included with the herbicide.

Rye: late tillering stage

Use an eyespot fungicide if more than 20% of the shoots bear eyespot lesions at the late tillering to first node stage. If regular usage of MBC fungicides (benomyl, carbendazim or thiophanate methyl) in previous years suggests that MBC-resistant strains of the eyespot fungus are likely to have been selected, then such materials should only be used in combination with a fungicide active against eyespot but having a different mode of action.

If one or more leaf diseases (mildew, brown rust, rhynchosporium leaf blotch or septoria leaf spot) is obvious and generally distributed in the crop between the beginning of spring growth and the emergence of flag leaf, a single application of a broad-spectrum fungicide should be made.

TABLE 6.9 Foliage-applied fungicides: major disease control recommendations

Active ingredient(s)	Wheat					Barley						Oat		Rye	
	eyespot	mildew	septoria	brown rust	yellow rust	eyespot	mildew	rhynchosporiuim	brown rust	yellow rust	net blotch	crown rust	mildew	eyespot	mildew
benodanil				√	√				√	√					
benomyl	√*	(√)	(√)			√*	(√)	√						√*	(√)
captafol			√					√							
carbendazim	(√*)	(√)	(√)			√*	(√)	√						√*	(√)
carbendazim + "activator"	√*	(√)	(√)			√*	(√)	√							
carbendazim + chlorothalonil	w √*	w (√)	w √			√*	(√)	√							
carbendazim + dithiocarbamate[1]	(√*)	(√)	(√)	(√)	(√)	(√*)	(√)	√	(√)	(√)					
carbendazim + maneb + sulphur	(√*)	(√)	(√)	(√)	(√)	w (√*)	w (√)	w (√)	w (√)	w (√)					
chlorothalonil			w √					√							
ethirimol + captafol		√	√				√	√							
fenpropimorph		√		√	√		√	(√)	√	√		(√)	√		
fenpropimorph + iprodione[2]		√	√				√	(√)	√	√	√				
flutriafol + captafol	w (√)	w √	w √	w √	w √										
flutriafol + carbendazim	w √*	w √	w √	w √	w √	√*	√	√	√	√	√				
iprodione[2]			√								√				
manganese + zinc dithiocarbamate			(√)		(√)										

nuarimol		(√)					√						√		√
prochloraz	√	(√)	√			√	(√)	√			√			√	(√)
prochloraz + carbendazim	√	(√)	√			w	w	w			w			√*	(√)
						√	(√)	√			√				
propiconazole	(√)	√	√	√		(√)	√	√	√	√	√		√		√
propiconazole + carbendazim	w	w	w	w		w	w	w	w	w	w				
	√*	√	√	√		√*	√	√	√	√	√				
propiconazole + tridemorph	(√)	√	√	√	√	(√)	√	√	√	√	√				
propineb					(√)										
pyrazophos							√	(√)			√				
sulphur		(√)													
thiophanate-methyl	√*	(√)	(√)											√*	(√)
triadimenol		√		√	√		√	√	√	√		(√)	√		√
triadimenol + carbendazim	w	w	w	w	w	w	w	w	w	w				√*	√
	√*	√	(√)	√	√	√*	√	√	√	√					
triadimefon		√		√	√		√	√	√	√		(√)	√		√
triadimefon + captafol		√	√	√	√										
triadimefon + carbendazim	w	w	w	w	w	w	w	w	w	w				√*	√
	√*	√	(√)	√	√	√*	√	√	√	√					
tridemorph		w			w		√			(√)			√		
		(√)			(√)										
tridemorph + carbendazim + maneb	√*	(√)	(√)	(√)	(√)	√*	√	√	(√)	(√)	(√)				
triforine							√								

[1]Some products do not include recommendation for spring wheat.
[2]Control of *Septoria tritici* is only partial.

*MBC resistant strains are not controlled by benomyl, carbendazim or thiophanate-methyl.
√ indicates disease controlled.
(√) indicates partial control.
w indicates winter wheat/barley only.

Two Nodes Detectable to Flag Leaf Emerged Period

Wheat: second node detectable to flag just visible

Fungicidal sprays may occasionally be necessary during this period especially if no spray was applied at the early jointing stage, if yellow rust appears in susceptible varieties (NIAB rating 3 or less) or if mildew is well established and active, covering 10% or more of the leaf area on the oldest green leaves. If a fungicide is applied at "flag leaf just visible" stage and the weather is wet consider the use of a broad-spectrum material.

Oats: stem extension to panicle emergence

Appropriate fungicides should be applied if the following criteria are met:–

Mildew. Spray if it becomes obvious on the third youngest leaf.

Crown rust. Spray if this disease is found anywhere on the top 3 leaves of susceptible varieties (NIAB disease rating of 4 or less). Spray more resistant varieties if the disease shows signs of spreading on the upper leaves.

Rye: up until flag leaf emergence

Apply a single application of a broad-spectrum fungicide if one or more leaf diseases is obvious throughout the crop. (Mildew, brown rust, rhynchosporium leaf blotch or septoria leaf spot.)

Barley: flag leaf emergence to full ear emergence

Diseases building up on the upper leaves at this time (and in the post-flowering period) can accelerate senescence and reduce grain size. Such pathological acceleration of leaf death can be of particular importance on well structured, moisture retentive soils where leaf area duration is less likely to be limited by moisture stress.

Mildew, rhynchosporium and brown rust. Susceptible varieties on medium to heavy soils should be sprayed if these diseases develop on the flag or second leaves.

Yellow rust. Spray if there is evidence of spread from existing foci or if the disease is found to be generally distributed in the field (10 affected tillers per square metre).

Net blotch. Spray if the disease is readily found on any of the top 3 leaves.

Wheat: flag leaf emergence

If a spray was applied at "flag leaf just visible" no further application should be made for at least 14 days. Fungicides applied even earlier in the season will slow down development of disease epidemics and can reduce the need to spray at flag leaf emergence

If a spray is applied to control any disease at flag leaf emergence it is advisable that a broad-spectrum fungicide should be used to give protection against all pathogens present (even at trace levels) within the crop.

Septoria. Spray if the disease is obvious on the lower leaves or if frequent leaf wetness (at least 1 mm rain on 4 days in the previous 2 weeks) provides conditions conducive to its spread.

Mildew. Spray if the disease is obvious on the oldest green leaves.

Yellow rust. With highly susceptible varieties, spray at first sign of rust. Spray the more resistant varieties (NIAB rating 4 or more) if the disease shows signs of spreading from existing foci or if the disease is found to be generally distributed through the field (10 tillers or more affected per square metre).

Brown rust. Spray if the disease can be found even at trace levels in crops of susceptible varieties.

Ear Emergence to Harvest Period

Wheat: early ear emergence to anthesis

If a spray was applied between flag leaf emergence and booting, no further application should be made for at least 14 days.

If mildew, yellow rust, brown rust or septoria are present or if, in the absence of visible disease, it is known that there is high risk of septoria (at least 1 mm rain for each of 4 days in the previous 2 weeks) or of brown rust (variety with a NIAB resistance score of 2 or less) then a broad-spectrum fungicide should be applied, the material chosen being one known to have particular activity against the diseases from which the crop is at risk.

If little or no disease is present, and the risk criteria outlined above do not apply, the use of a prophylactic spray of a broad-spectrum fungicide may be considered. Such a spray must be regarded as an insurance policy against the major diseases which frequently (though not invariably) build up after flowering on flag leaves and glumes, and also against the minor parasites which often colonise the aging leaves and accelerate their senescence. The decision as to whether or not to take protective insurance action rests on three considerations: the value of the item to be protected, the cost of protection and the assessment of risk.

Investment in insurance spraying is most justified in crops with a high yield potential, especially those destined for seed or milling and those crops which have received liberal applications of nitrogen rendering them more susceptible to leaf diseases.

The cost of protection includes: the cost of fungicide, application costs and damage due to wheelings. In a previously unwheeled crop the latter will amount to a 3% loss if a 12 m spray boom is used. If use can be made of existing tractor lanes the losses will be smaller.

Risk assessment requires consideration of the following points:–

- Susceptiblility of variety to major pathogens (NIAB resistance scores)
- Responsiveness of the variety to fungicide application; given in NIAB Farmers Leaflet No. 8 *Recommended varieties of cereals,* (National Institute of Agricultural Botany, Cambridge).
- The presence of disease in the immediate neighbourhood (Advisory Services' disease intelligence schemes and telephone information services may be helpful).
- Knowledge of local conditions. Farmers will know from past experience if their fields are subject to particular disease risks. Coastal areas and valley bottoms where mist hangs are, for example, particularly prone to severe attacks of septoria.
- Soil conditions and root diseases. The active life of a wheat leaf may be limited either by moisture stress, caused by high soil moisture deficit, by soil structural problems restricting rooting, or by root disease reducing water uptake in the soil; or by the activity of parasitic fungi, either major pathogens or the late invaders of aging tissue. On a well structured, moisture retentive soil, moisture stress is unlikely to be the limiting factor in determining leaf longevity except in very dry seasons or in situations where root diseases are a problem. Under such circumstances the role of late developing diseases is likely to be of greater importance and action to control them is more likely to be economic. On thin, light, drought-prone soils on the other hand the late application of fungicides at this time will only rarely have much effect on the rate of senescence which will be determined mainly by moisture stress.

Most wheat crop situations will justify the application of fungicide at this stage if no spray was applied at flag leaf emergence.

Barley: full ear emergence to end of flowering

Spray applications of the winter barley are seldom necessary after full ear emergence, but appropriate fungicides may be applied up to the end of flowering if net blotch or brown rust are found on the top three leaves of susceptible varieties.

Wheat: watery ripe stage

Do not treat crops until at least 14 days after a previous spray.

Rusts. On susceptible varieties (NIAB score 3 or less) spray if yellow or brown rusts can readily be found on the upper leaves.

Minor Pathogens. If the flowering period has been damp and overcast, and especially if no fungicide has been applied at ear emergence, minor parasitic pathogens can build up and accelerate leaf senescence. This is likely to be of particular significance if cool cloudy weather prolongs the grain fill period. In such years the application of an appropriate fungicide after the end of flowering often prolongs active leaf life and gives an economical increase in yield.

Latest time for safe application of fungicides

The harvest interval established for the safe use of fungicides used on wheat and barley limits the late use of a number of materials. The latest recommended application times shown in Table 6.10 do not define the safe harvest interval which is stated in terms of the number of days or weeks which should elapse between application and harvest on the labels of products concerned. In normal seasons the recommended latest application time will allow an adequate harvest interval.

Recommendations for Control of Specific Diseases

Take-all

There is no reliable chemical control for take-all (*Gaeumannomyces graminis*). Seed treatment with triadimenol + fuberidazole sometimes gives a marked suppression of take-all but this effect is not consistent. Cultural control has traditionally been based on rotation but in recent years more and more farmers have exploited the take-all decline phenomenon by growing continuous cereals. Since barley although susceptible suffers less seriously from disease than does wheat, barley crops may be taken during the peak take-all years and the field returned to wheat in year 5 or 6 of the sequence. During the peak take-all years losses may be minimised, at least on well structured heavy soils, by strict application of the principles of good husbandry. A field must be well drained, any soil structural problem corrected and a balanced fertiliser regime maintained. In a high-risk situation aim for a phosphate index of 3. Since loose puffy soils favour the spread of the fungus over the roots care should be taken to achieve a good firm seedbed. Early sowing increases the risk of severe take-all losses. It is advisible to sow first wheats first and then long-term wheats (i.e. fields already in the take-all decline phase), leaving the high risk, second, third and fourth wheat

TABLE 6.10 Latest recommended time for fungicide use in wheat and barley

	Before ear emergence	Ear just visible	Ear emergence	Start of flowering	Grain watery ripe	Weeks prior to harvest 6	5	4	3	2	1	Harvest
benodanil			●									
benomyl												●
captafol (on barley)	●											
captafol (on wheat)			●									
carbendazim (on barley)			●									
carbendazim + "activator" (on barley)		●										
carbendazim ± "activator" (on wheat)												●
carbendazim + maneb or mancozeb											●	
carbendazim + maneb and mancozeb (on wheat)			●									
carbendazim + maneb + sulphur			●									
chlorothalonil ± carbendazim			●									
fenpropimorph							●					
fenpropimorph + iprodione					●							
flutriafol + captafol (on wheat)			●									
flutriafol + carbendazim					●							
iprodine (on barley)			●									
iprodine (on wheat)					●							
manganese/zinc dithiocarbamate (on wheat)			●									

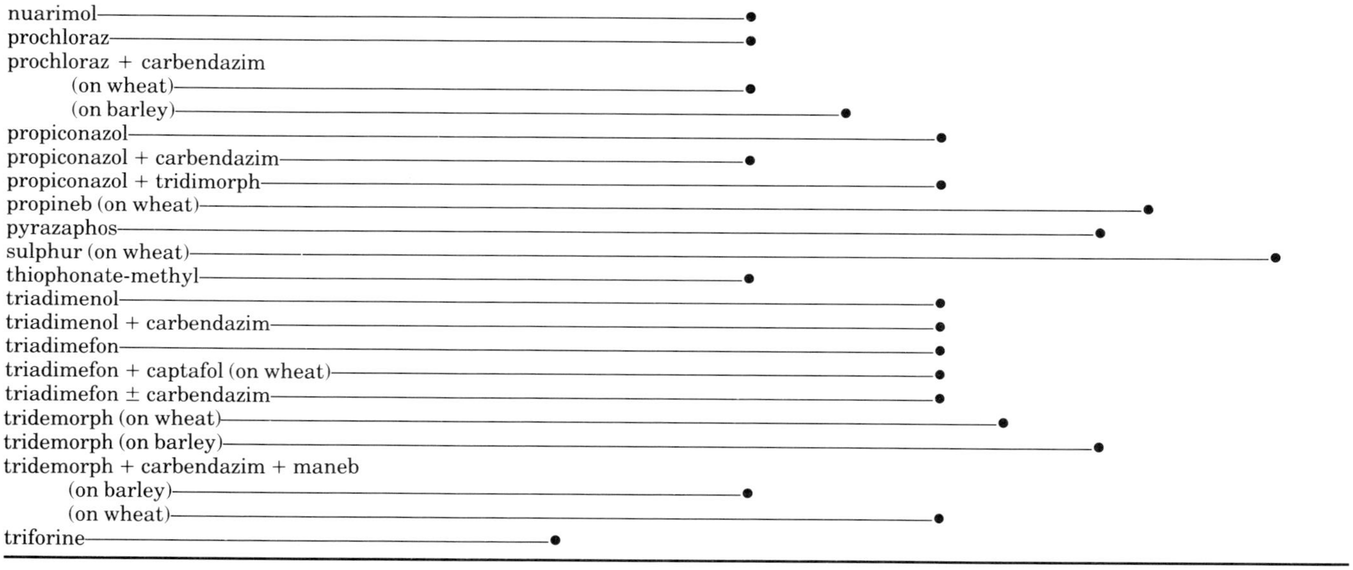
nuarimol
prochloraz
prochloraz + carbendazim
(on wheat)
(on barley)
propiconazol
propiconazol + carbendazim
propiconazol + tridimorph
propineb (on wheat)
pyrazaphos
sulphur (on wheat)
thiophonate-methyl
triadimenol
triadimenol + carbendazim
triadimefon
triadimefon + captafol (on wheat)
triadimefon ± carbendazim
tridemorph (on wheat)
tridemorph (on barley)
tridemorph + carbendazim + maneb
(on barley)
(on wheat)
triforine

crops until last. There are no significant differences in susceptibility between varieties, but some cultivars (notably Maris Huntsman) appear able to withstand infection without experiencing the degree of yield loss suffered by other varieties.

Eyespot

Rotational control of eyespot (*Pseudocercosporella herpotrichoides*) normally requires a non-cereal break of at least two years. A one year break can, however, reduce levels of infection if the stubble is ploughed down before the break and not brought to the surface after it, by using direct drilling or minimum cultivations for establishing the subsequent cereal crop. Alternatively, stubble left on the surface before the break can be ploughed under after it, for example where an oilseed rape crop is direct drilled into cereal stubble and the rape stubble is ploughed before the next crop. Varieties differ considerably in their susceptibility to eyespot and the more susceptible ones should not be grown in high risk situations. Excessively early sowing should be avoided, as should excessive nitrogen usage which renders infected crops more prone to lodging.

Growth regulators will not control disease but, by strengthening the stems, they may reduce the risk of the lodging which is the most important cause of yield loss in infected crops. Chemical control of the disease was until recently based on the use of benzimidazole (MBC) fungicides. In recent years, however, MBC resistant strains of the eyespot fungus have been found to be widespread in the major cereal growing areas. Such strains are particularly common in fields where there has been routine use of such fungicides over a number of years. The "BWR-pathotype" of the fungus (which attacks barley, wheat and rye) produces resistant strains far more regularly than does the "BW-pathotype" (which attacks barley and wheat only). Prochloraz has been found to give effective control of the MBC resistant strains of the fungus. Propiconazole and flutriafol also show some activity against the eyespot fungus, and triadimefon which has some protectant activity can, when applied in the autumn, reduce levels of the disease overwintering in the crop.

It is normally recommended that an eyespot spray should be applied if 20% of the shoots bare lesions at the leaf sheath erect to first node visible growth stage. For most crops this means that the spray will be applied in April and by the time the effects of the chemical have worn off temperatures will usually be too high for further infection to occur.

Sharp eyespot

Although some fungicides are known to show activity against the sharp eyespot fungus (*Rhizoctonia cerealis*), attempts to control the disease by chemical means have given erratic and generally disappointing results. At the moment there are no recommendations for the use of fungicides against this disease. Varieties are known to differ in their susceptibility to sharp eyespot but at the moment insufficient data is available to allow resistance scores for this disease to be given in the NIAB Recommended List of cereal varieties.

In fields known to have a history of sharp eyespot very early sowing of winter cereals should be avoided.

Seedling blight, brown foot-rot and ear blight

All the commonly used seed treatments give at least partial control of the seed-borne *Fusarium* spp. infection which is the main cause of seedling blight. Fungicidal control of brown foot-rot and ear blight is unreliable but sprays applied for the control of eyespot will reduce brown foot-rot systems and MBC + dithiocarbamate fungicides applied at about the time of anthesis will give at least partial control of ear blight symptoms.

The effects of brown foot-rot can be minimised by good husbandry, in particular by good drainage and avoidance of soil structural problems.

Snow rot

The organism causing snow rot (*Typhula incarnata*) is a member of the "Basidiomycotina" group of fungi which includes the rusts. Fungicides active against the rusts will also afford some protection against snow rot but for best results they need to be applied as protectants before the symptoms appear and in most parts of the country the erratic occurrence of the disease argues against routine precautionary treatment. However, in light land areas (especially in Scotland) where snow rot is known frequently to present problems its possible development in the late winter may be one of the factors influencing a decision on fungicide usage in the autumn. There appear to be varietal differences in susceptibility to snow rot. For example the barley varieties Gerbel and Pirate appear to be more resistant to the disease than Igri or Panda. At present there is not enough information to allow firm resistance scores to be given.

Although all cereal species are vulnerable to snow rot, losses are most commonly encountered in winter barley.

Septoria leaf spot and glume blotch

None of the currently grown wheat varieties shows good resistance to both septoria leaf spot (*S. nodorum*) and glume blotch (*S. tritici*). The NIAB Recommended List of cereal varieties does however indicate some varietal differences in susceptibility and the more highly susceptible cultivars should be avoided in high risk areas: regions of high rainfall, misty valleys or coastal areas subject to sea mists.

Most of the currently available fungicide seed treatments will reduce, though not completely eliminate, seed-borne infection by *S. nodorum* and triadimenol + fuberidazole seed treatment will give some suppression of *S. tritici* during the early stages of crop growth. Whether or not seed treatment was used however, a fungicidal spray may sometimes be necessary against *S. tritici* in the late winter or early spring, and *S. nodorum* is one of that complex of pathogens against which broad-spectrum fungicides can be applied between flag leaf and ear emergence. Prochloraz, propiconazole and chlorothalonil show good activity against both pathogens. Some suppression of both diseases may also be achieved by the use of fungicide formulations and mixtures based on benzimidazole (MBC) materials plus dithiocarbamates but strains of *S. tritici* resistant to MBC fungicides are now widespread. Alternatively triadimefon also gives good control of *S. tritici*. Captafol is very effective against *S. nodorum*. These fungicides need to be applied before the disease becomes established on the upper leaves.

Leaf blotch of barley

Varietal resistance offers an effective method of controlling leaf blotch of barley (*Rhynchosporium secalis*) but may need to be augmented by application of an effective fungicide. Captafol, chlorothalonil, prochloraz, propiconazole, tridimefon and the benzimidazole fungicides (alone or in combination with other materials) all give effective control of this disease.

Net blotch of barley

The increase in net blotch (*Pyrenophora teres*) in the late 1970s was associated with the widespread use of the very susceptible variety Sonja. Most modern varieties are far less susceptible to the disease but the use of fungicidal sprays to control it may be necessary in wet seasons. Prochloraz and propiconazole are very effective against the disease. Partial control may also be achieved by the use of iprodione and by combinations of tridemorph or triforine with dithiocarbamate fungicides. Most of the currently used fungicidal seed treatments, including organomercury, will give some control of the seed-borne phase of the disease.

Leaf stripe of barley

All currently used fungicide seed treatments give good control of leaf stripe (*Pyrenophora graminea*) on spring barley but it should be noted that materials based on nuarimol and on tridimenol + fuberidazole (without the addition of imazalil) will give only partial control in the autumn-sown crop.

Cephalosporium leaf stripe

Where severe attacks of leaf stripe (*Cephalosporium gramineum*) have occurred a non-graminaceous break crop of at least one year is recommended. The next cereal crop should be sown after ploughing rather than by direct drilling and where the disease has occurred straw incorporation should be avoided.

Powdery mildew

A different form ("forma specialis") of the mildew fungus (*Erysiphe graminis*) occurs on each cereal species so infection between wheat and barley, for example, does not take place.

Because mildew spores can be carried considerable distances by the wind, hygiene practices aimed at reducing inoculum levels can never be more than partially effective. However, severe infection at the seedling stage is most likely to occur in crops close to strong sources of inocolum and it is advisable therefore to plough stubbles with infected volunteers, or spray with suitable herbicide, before nearby early-sown crops of the same species emerge.

Wheat mildew in particular is often especially severe in certain fields such as those on rich organic soils or soils subject to manganese deficiency. It is advisable not to grow highly susceptible varieties in such a situation. However, even a variety with a high score for resistance can sometimes succumb to a race of the fungus which is well adapted to it, and the existence of such physiologic races makes varietal selection difficult. The NIAB Recommended List groups winter wheat, though not other winter cereal, varieties into Diversification Groups; the various groups having their mildew resistance based on different genes or groups of genes. Selection of two or more varieties from different Diversification Groups will reduce field to field spread of the disease. If the different varieties are grown as a blend the level of disease in this crop will be lower than the average levels in crops of the same varieties grown separately.

Mildew is relatively easy to control chemically and a wide range of fungicides active against the disease are available for application both as seed treatments and sprays. In winter barley it may be necessary to apply mildew fungicides in autumn, spring and summer. In winter wheat it is seldom necessary to apply more than two sprays to achieve good control of the disease in spring and summer. Where more than one treatment is required it is advisable to use chemicals with different modes of action (either in sequence or in mixtures) to reduce the risk of selecting out fungicide resistant strains of the pathogen.

Brown rust of wheat

Wheat varieties differ in susceptibility to brown rust (*Puccinia recondita*) and in recent years some initially quite resistant cultivars have succumbed to previously undetected races of pathogen. Varieties known to be highly susceptible (NIAB rating 3 or less) are best avoided in areas prone to the disease but if they are grown in such areas they should receive a precautionary spray as soon as the flag leaf has emerged with a second spray being applied after ear emergence (2 or 3 weeks after the first). In areas where the disease risk is lower susceptible varieties should be sprayed as soon as brown rust is seen on the upper leaves. More resistant varieties should receive a fungicide application if brown rust can be readily found on the upper leaves and there is evidence that the disease is spreading. Spraying after the watery ripe stage is unlikely to be cost effective

Good control of wheat brown rust can be obtained using the fungicides fenpropimorph flutriafol, propiconazole, triadimefon and triadimenol. Dithiocarbamate fungicides (either alone or in combination with other materials such as tridemorph) have a protectant action against the disease and may check its development if the attack is not severe

Brown rust of barley

The currently grown winter barley varieties show considerable differences in their susceptibility to brown rust (*Puccinia hordei*) though none is completely resistant to the disease.

While brown rust may often be found in the late autumn and winter, winter barley crops generally grow away from the disease in the spring and it is seldom necessary to apply a spray until flag leaf emergence. Susceptible varieties (NIAB rating 4 or less) should be sprayed if the disease can readily be found on any of the top three leaves. The fungicides fenpropimorph, flutriafol, propiconazole, triadimefon and triadimenol will give good control of the disease, and moderate attacks may be checked by the protectant action of dithiocarbamate fungicides applied either alone or in combination with other materials such as tridemorph.

Yellow rust of wheat

The most severe attacks of this disease (*Puccinia striiformis* f. sp. *tritici*) occur in crops where infection has been established in the autumn, so stubble hygiene to destroy infected volunteer plants before surrounding autumn sown crops emerge is very important. Varieties differ widely in their susceptibility to this disease and, as with wheat mildew (see page 151) it is possible to reduce the risks of field to field spread by choosing two or more varieties with different genetic bases for their resistance to the disease (see the current NIAB Farmers leaflet No. 8), *Recommended varieties of cereals* or ADAS Booklet 2257, *Use of fungicides and insecticides on cereals.* Varieties from different Diversification Groups may be grown as blends. It is possible to choose varieties which are in different Diversification Groups with respect to both yellow rust and mildew, but the choice here is much more restricted than if one seeks to diversify for only one of the two diseases. If it is desired to grow only one variety on the farm then it is a wise precaution to grow a variety from a different Diversification Group in the following year.

Highly susceptible varieties (NIAB rating 4 or less) should be sprayed as soon as the disease appears in the crop. Further sprays may be necessary if the first is applied very early in the season. Moderately susceptible varieties (NIAB rating 5 or 6) should be sprayed if there is evidence that the disease is spreading from early established foci of infection, or if it is more generally distributed through the crop (10 tillers or more affected per square metre). Less susceptible varieties are unlikely to require treatment unless cool (10 to 15°C) and moist conditions persist. Sprays applied after the watery ripe stage are unlikely to cost effective. Yellow rust is well controlled by fenpropimorph, flutriafol, propiconazole, triadimefon, tridimenol and by tridemorph in combination with a dithiocarbamate. Dithiocarbamates alone have a protectant action against the disease.

Yellow rust of winter barley

Winter barley varieties differ widely in their susceptibility to yellow rust (*Puccinia striiformis* f. sp. *hordei*) but at the moment there is no diversification scheme for barley as there is for wheat. While the disease may be found from late autumn onwards spraying before growth starts in the spring is unlikely to be worthwhile. Thereafter susceptible varieties (NIAB resistance 3 or less) should be sprayed if the disease spreads onto the new spring growth or if it subsequently appears up until the time of ear emergence. Less susceptible varieties should not be sprayed unless there is evidence of spread from early established foci of infection or if the disease is more generally distributed (10 tillers or more affected per square metre). A second application may be necessary if the first is given early in the season. Barley yellow rust may be controlled by the same range of fungicides recommended for the control of the disease on winter wheat.

Crown rust of oats

Since oats are less widely grown than are wheat and barley, stubble hygiene is likely to provide a particularly effective method of breaking the disease cycle of crown rust (*Puccinia coronata*) on those farms where the oat crop is grown. With yellow rust of wheat and barley there is a much greater likelihood that the disease will spread in from surrounding areas over which the farmer has no control). It is particularly important therefore that volunteer plants in oat stubble should be ploughed under or sprayed off before the earliest autumn sowings emerge.

Susceptible varieties (NIAB rating 4 or less) should be sprayed as soon as the disease appears in the crop. Other varieties need be sprayed only if the rust shows signs of spreading from early established foci. If crown rust can be readily found in the crop when a spray is being applied against mildew a fungicide active against both diseases should be used. Fenpropimorph and triadimefon are the only materials currently recommended for use against crown rust.

Late season diseases of leaf and ear

While the upper leaves of all cereal species can suffer from attacks by minor pathogens (*Botrytis cinerea*, *Alternaria alternata* and others) after anthesis, the new large-eared varieties of wheat seem to suffer particularly severely from their effects in hastening the aging of green photosynthetic tissue. If no fungicide was applied at ear emergence a broad-spectrum material should be applied if warm humid weather occurs around the time of anthesis. The use of formulated or tank-mixed mixtures of MBC dithiocarbamate fungicides are especially recommended for those varieties known to be highly susceptible to Fusarium ear blight.

Loose and covered smuts and bunt

See note on treatment against seed-borne diseases on page 134.

Ergot

The resting bodies of ergot (*Claviceps purpurea*) seldom survive in the soil for more than a year in the absence of grass weeds so rotation offers an effective method of control. If a cereal is to follow an affected crop the land should be ploughed so that shed resting bodies (sclerotia) are buried too deeply for the stalks to reach the surface when they germinate the following spring. So far as wheat is concerned the control of black-grass is very important as it deprives the pathogen of an alternative host on which it can bulk up within the crop. If black-grass or meadow foxtail flourish on the headland or field margins they should be topped before coming into flower. Using normal seed cleaning techniques it is very difficult to get rid of ergots from a seed sample and it is best therefore not save seed from infected crops.

"Open-flowering" varieties are particularly susceptible to infection by *Claviceps purpurea* and durum wheat can therefore suffer particularly badly. Although at the time of writing no chemical has a label recommendation for the control of ergot it is known that MBC fungicides show activity against the disease and some growers of durum wheat have applied these chemicals as precautionary sprays at the time of anthesis.

Barley yellow dwarf virus (BYDV)

Stubble hygiene is important in the control of BYDV; severe early attacks may occur if infected wheat grasses and volunteer cereals are not destroyed before the next crop is sown. The earlier a crop is sown the greater is the risk of infection and in coastal and estuarine areas of southern England and Wales, and in sheltered areas close to grassland elsewhere in the country, very early drilling, especially of winter barley, should be avoided.

In "virus-prone" areas it is advised that an aphicide should be applied in late October or early November to crops sown before mid October if BYDV is commonly a problem on the farm or in the vicinity. Elsewhere in the country earlier sown crops need be sprayed only if forecasts suggest that spread of the virus is likely, or if previous experience suggests that the situation on a particular field is especially conducive to the development of the disease. The BYDV forecasts, based on the numbers of aphids flying in the autumn and the percentage able to transmit the virus, are issued in late October by ADAS disease intelligence schemes and Telephone Information Services.

Barley yellow mosaic virus (BaYMV)

If BaYMV appears on a field the application of very strict hygiene measures to limit the transport of infected soil can sometimes contain the problem and prevent the spread to other fields on the farm. Where possible access to the affected field should be restricted to that necessary for essential husbandry and any equipment used on it should be hosed down to remove contaminated soil before the equipment is used in other fields. Spread within the affected field can be reduced if the land is ploughed rather than cultivated with rigid tine equipment.

It is safest not to grow winter barley on affected land for some years. Spring-sown crops do not develop symptoms though they may act as carriers of the virus. It is not known how long a break from susceptible crops would be necessary to eliminate the pathogen.

If winter barley has to be grown on an affected field advice should be sought on choice of variety. There are considerable varietal differences in susceptibility to the disease. Since, however, strains of the virus exist which differ in their virulence to different varieties, the choice of variety to grow in an infected field will depend on which variety was previously grown in that field.

PEST CONTROL

Between-Crops and Sowing Period

Cultural methods

Cultivations

Plough grass fields that are to be sown to cereals in time to leave a period of at least four weeks before sowing to reduce aphid transmission of BYDV and remove egg-laying sites for frit fly (*Oscinella frit*), crane fly (various species, typically *Tipula paludosa*), swift moth (*Hepialus* spp.) and crambid moth (*Agriphila straminella*).

Grass weeds and volunteer cereals in stubble can also carry over BYDV-infected aphids and attract egg-laying pests. Cultivations alone may be sufficient to control these weeds and, if carried out so as to keep soil bare for at least four weeks before sowing, reduce the pest risk.

Within the limits set by soil type and condition, produce firm fine seed beds to reduce slug (usually *Deroceras reticulatum*) movement especially on heavier soils and where wheat is drilled after grass, clover, oilseed rape, maize, peas or beans. Firm, fine seedbeds also reduce wireworm (*Agriotes* spp.) damage where the pest is present.

Shallow sowing reduces the damage caused by wheat bulb fly larvae (*Delia coarctata*) feeding. Avoid overdeep sowing, which increases the severity of the damage caused by most soil pests.

Crop residues

Crop residues in the soil can greatly encourage slug activity. Any process which assists the rapid decay of the residues helps to reduce slug problems. Chop cereal straw, oilseed rape, pea and bean residues as finely as possible before incorporation.

Stubble burning is a useful hygiene measure against a number of cereal pests.

Direct drilling

In medium and heavy soils, direct drilling reduces the risk of attack by some stem-boring pests and favours beneficial organisms such as aphid predators and earthworms. However, it increases the risk of field slug damage and frit fly damage which may require chemical treatment.

Time of sowing

Sowing wheat and barley before mid-October increases the risk of attack by frit fly, yellow cereal fly (*Opomyza florum*), gout fly (*Chlorops pumilionis*), cereal aphids and BYDV infection. Late sowing, after the end of October, reduces the risk of these pests but increases the damage done by wheat bulb fly and slugs.

Cereal species and variety

Stem nematode. Avoid sowing rye on land known to be infested with stem nematode (*Ditylenchus dipsaci*) and use only resistant varieties of oats in this situation; barley and wheat are not affected.

Cereal cyst nematode. Where cereal cyst nematode (*Heterodera avenae*) is present, avoid growing oats unless a satisfactory resistant variety is available. Wheat, barley and rye are less susceptible to attack, in that order. Varietal resistance has been bred into some varieties of spring barley.

Wheat bulb fly. Winter oats are immune to this pest which will attack winter barley and rye (for which no insecticidal treatments are available) as well as winter wheat.

Wireworms. Barley and rye are less susceptible than oats and wheat.

Chemicals

Seed treatments

Wheat bulb fly. Seed treatment to give some protection to wheat (only) against this pest is recommended for crops sown late after potatoes, sugar beet and other 'open land' situations if damage is forecast. Products based on carbofenothion and chlorfenvinphos are available.

Wireworm. Gamma-HCH seed treatment with or without organomercury will give adequate protection to cereals against low to moderate populations of wireworm. Fonofos applied for wheat bulb fly control will also give some protection against wireworm. Gamma-HCH treatment must not be applied to seed together with wheat bulb fly treatment. Only treat undamaged seed with a moisture content of 16% or less and sow treated seed as soon as possible to avoid injuring it. Large populations of wireworm require an overall incorporated insecticide application for adequate control (see p. 158).

Protection of wildlife. Seed treated with insecticide is dangerous to birds and should be well covered with soil at sowing. However, shallow drilling makes the most effective use of seed treatment so a level seedbed is required. Collect any spilt treated seed and dispose of it where birds will not take it. Avoid drilling insecticide-treated seed after December when birds are more likely to seek it out.

Insecticidal soil treatments

Insecticidal sprays and granules are normally applied only when damage is seen. When large numbers of leatherjackets, wheat bulb fly or wireworm are known to be present applications at sowing time may be justified. Treatments are shown in Table 6.11.

Gamma-HCH as a spray or bait should not be used where potatoes or carrots are to follow within 18 months because of the risk of taint.

Sprays and granules are convenient for treating large areas; baits, which use less insecticide, are useful for treating isolated patches. Proprietary gamma-HCH bait pellets are available and it is possible to prepare baits using gamma-HCH or fenitrothion and bran. Thorough and even mixing is essential using just enough water to moisten the bran without causing it to stick in lumps.

TABLE 6.11 Insecticidal soil treatments for use at sowing time

	Pest		
Treatment	Wheat bulb fly	Wire-worm	Leather-jackets
Chlorfenvinphos			
Spray just before or after sowing	√		
Chlorpyrifos			
Spray or granules applied after sowing (delay use of granules against wheat bulb fly until early winter just before egg hatch)	√		√
Fenitrothion			
Bait application after sowing			√
Fonofos			
Spray or granules at or just before sowing	√	(√)	
Gamma-HCH			
Bait: application a few days before sowing best			√
Spray: worked into seedbed		√	√

Molluscicides

Slug bait pellets based on metaldehyde or methiocarb may be applied before ploughing stubbles; they may be drilled with the seed or be applied immediately after drilling. Re-treatment may be required to prolong protection. Methiocarb applied for slug control will give some control of leatherjackets.

Weed control chemicals

Herbicides used against grass weeds and volunteer cereals in this period have an indirect effect in the control of a number of cereal pests especially the egg-laying 'ley pests' and aphids. They are a useful adjunct to cultivations for post-harvest hygiene measures.

Emergence to Tillering (Winter) Period

Except where soil pests are known to be present in large numbers it is recommended that insecticides are applied to cereals only when damage is seen. Treatments applied in this way are indicated in the section below dealing with specific pests (p. 160). Treatment that can be applied in this period in anticipation of damage are:–

Aphids – BYDV. In high risk BYDV-prone areas spray cereal seedlings in late October or early November using an approved synthetic pyrethroid, organophosphate or carbamate aphicide (p. 90). Only spray aphicide on cereals outside high risk areas when the crop is early-sown and forecasts suggest that BYDV spread is likely, or when there is local knowledge that BYDV often infects a particular field.

Wheat bulb fly. Treatments are only available for wheat, not barley or rye (oats are immune). On mineral soils chlorpyriphos granules may be appplied in January before eggs begin to hatch. Spray treatments which can be applied at or immediately before egg-hatch are given, together with preferences for different soil types in the specific pest section below (p. 160).

Frit fly. Where damage is anticipated apply a spray of chlorpyrifos at the time of plant emergence.

Yellow cereal fly. Use one of the sprays given in the specific pest section below (p. 162) before or at egg-hatch. ADAS warnings of exceptional egg numbers and time of egg-hatch are available locally.

Tillering to Jointing (Spring) Period

Cultural measures

During this period the provision of adequate nitrogen fertilizer will enable crops to recover from attacks of pests such as wheat bulb fly, frit fly, leatherjackets, swift moth and crambid moth larvae. Where soil conditions permit, rolling to consolidate the soil can also assist recovery from damage and reduce the activity of field slugs.

Chemical treatments

Wheat bulb fly. Spray treatment to combat tiller damage (the 'deadheart' stage) caused by continued feeding by larvae can be applied in this period (see below).

Leatherjackets. Sprays or granules of chlorpyrifos; sprays of gamma-HCH, quinalphos or triazophos; or baits with fenitrothion or gamma-HCH can be applied. They are most effective when the soil is damp and the temperature is above 6 °C. Because of the risk of taint, gamma-HCH should not be applied where potatoes or carrots are to follow within 18 months.

Stem Elongation to Harvest Period

Chemical treatments

Avoid unnecessary applications of insecticides by making thorough estimates of the numbers of the pests present throughout the fields concerned and, using suggested threshold levels as a guide, spray only when advisory entomologists suggest the operation is worthwhile.

Cereal aphids. Apply a single spray of aphicide if damage threshold levels are exceeded (see p. 162).

Wheat blossom midges (*Contarina tritici* and *Sitodiplosis mosellana*). Spray with chlorpyrifos, fenitrothion (which can be applied from the air) or triazophos between the time of ear emergence and flowering (GS 57–61) if two or more midges are seen on emerging ears of wheat Further details of this pest are given in ADAS Leaflet 788.

Saddle gall midge. (*Haplodiplosis marginata*). Spray fenitrothion in June if a warning is given. The treatment is best applied 10 days after the first adults or egg masses are seen unless large numbers of egg masses are present when they are first detected, in which case the spray should be applied without delay. Further details of this pest are given in ADAS Leaflet 657.

Thrips (*Limothrips cerealium* and *L. denticornis*). Spraying with fenitrothion during the later stages of ear emergence (GS 55–61) is likely to be worthwhile if two or more adult thrips can be found during the early stages of ear emergence.

Recommendations for Control of Specific Major Pests

Wheat bulb fly (*Delia coarctata*)

Sow winter wheat or winter barley early, before the end of October. When very heavy attacks are expected (e.g. after ADAS soil sampling has indicated large numbers of eggs present) consider oats as an immune alternative cereal. Work fallow land to a fairly fine tilth before sowing, and provide adequate fertiliser in the seedbed. Avoid deep or uneven drilling. Rolling and top-dressing with nitrogenous fertiliser aids recovery of affected crops.

Seed treatments of carbophenothion or chlorfenvinphos provide useful protection especially on wheat sown from November until the end of December. Carbophenothion must not be used in Scotland. Apply seed treatments to good quality seed with a moisture content lower than 16 per cent. Do not store treated seed in a damp place nor for longer than necessary. Avoid spillages, very shallow sowing or deep sowing of treated seed.

Granules of fonofos or sprays of chlorfenvinphos or fonofos applied at or immediately before sowing are effective on mineral soils but less so on organic soils. Sprays of chlorfenvinphos or chlorpyrifos applied after sowing give useful control on mineral soils only. Chlorfenvinphos or fonofos sprays can be applied before egg-hatch (January to March depending on season) on mineral soils.

At egg-hatch on both mineral and organic soils, a spray of chlorpyrifos, pirimiphos-methyl or triazophos can be used. For organic soils an egg-hatch spray of chlorfenvinphos or fonofos is preferred.

When larval damage is first noted (the 'deadheart' stage) use a foliar spray of dimethoate, omethoate plus fonofos on all soil types.

Multiple treatments, involving combinations of two or more of the above recommendations, are often cost-effective against severe attacks. Thus on organic soils a seed treatment, egg-hatch spray and 'deadheart' spray may give best results. Sprays of chlorfenvinphos, omethoate and triazophos are cleared for aerial application. Further details can be found in ADAS leaflet 177, *Wheat bulb fly*.

Cereal aphids/BYDV: autumn attacks

Cultivate the previous grass crop or stubble as early as possible or use an effective herbicide such as paraquat to kill grass weeds and volunteer cereal plants before seedbed preparation. Plough grass effectively so that the sward is completely buried. Cereals sown after mid-October usually suffer much less damage from BYDV than earlier sown crops.

Only use chemical treatment against the aphid vectors of BYDV for early-sown crops in areas and on farms which are virus-prone; early-sown crops outside the virus-prone areas in seasons when the risk of infection is great (note ADAS warnings); or crops which are infested with wingless aphids that have moved from the previous grass crop or weed-infested stubble.

The response of barley and oats to autumn aphicide treatment is usually greater than that of wheat which is normally less affected by the disease.

For the most effective control of BYDV use cypermethrin, deltamethrin, fenvalerate or permethrin spray in late October or early November; demeton-S-methyl is only slightly less effective for this purpose. In situations where very early infestations of bird-cherry aphid are present, either as progeny of early winged immigrant aphids or as wingless aphids resident within the field, an additional aphicidal spray in September or early October may be worthwhile.

Cereal aphids: summer attacks

Some varieties of winter wheat support lower numbers of the grain aphid and/or rose-grain aphid and may be less affected by direct feeding injury.

Examine crops from flag leaf emergence onwards. When five or more grain aphids are present on the ear, or if 30 or more rose-grain aphids infest the flag leaf and the weather is warm and settled apply single sprays of one of the following aphicides, using preferentially those such as phosalone and pirimicarb which have least deleterious effects on beneficial organisms such as bees: chlorpyrifos, demeton-S-methyl, dimethoate, heptenophos, oxydemeton-methyl, phosalone, pirimicarb, thiometon. Some of the above insecticides may be applied at a lower dose for grain aphid control after ear emergence. Treatments applied after the milky/cheesy ripe stage are of little or no value. See ADAS leaflet 586, *Cereal aphids*, for further details.

Yellow cereal fly (*Opomyza florum*)

Winter wheat sown before mid-October is more at risk than later sowings, but is often able to compensate for loss of tillers. Recovery from attack is aided by adequate seedbed fertiliser.

Insecticidal treatments are probably required only in crops suffering 30% or more loss of tillers. When necessary, use one of the following insecticides just before or at egg hatch (note ADAS warnings):– chlorfenvinphos, chlorpyrifos, cypermethrin, deltamethrin, fenvalerate, fonofos, omethoate, permethrin, triazophos. There is some evidence to show that synthetic pyrethroid insecticides applied in late October/early November against the aphid vectors of BYDV gave a noticeable residual effect against yellow cereal fly infestations. Sprays of chlorfenvinphos, omethoate and triazophos are cleared for aerial application.

Frit fly (*Oscinella frit*)

The risk of larval damage to winter cereals is greatly reduced by ploughing grass early and leaving an interval of at least four weeks before sowing. Direct-drilled cereals following desiccated grass swards are more susceptible to frit fly damage than crops sown after ploughing. Early sown cereals are much more at risk than those sown from late October onwards.

Apply one of the following insecticides at the first sign of larval attack, the 'deadheart' stage, when 10% or more plants are damaged or contain larvae: chlorpyrifos, fonofos, omethoate, triazophos. Note that treatment is of no value if applied when damage is obvious or extensive. Further details are given in ADAS Leaflet 110, *Frit fly*.

Field slug (*Deroceras reticulatum*)

Wheat or barley sown after oilseed rape, grass, maize, peas or beans on clay and silt soils are at risk to slug damage, especially after wet summer and autumn weather. Sow cereal crops in a reasonably fine and firm seedbed. Early-sown crops usually suffer less damage than those sown from late October onwards.

For crops known to be at risk or where test-baiting indicates much slug activity, apply pellets of metaldehyde or methiocarb bait before ploughing stubbles, a few days before sowing, mixed with the seed at sowing or immediately after sowing. Both metaldehyde and methiocarb are cleared for aerial application. Treatments are most effective if carried out when slugs are most active in moist, mild conditions. Test-baiting is done by placing small quantities of slug pellets under tiles, sacks or black polythene bags at about 10 points across the field. Examine the baits at about three-day intervals: when an average of four or more slugs per bait station is found, treatment is likely to be worthwhile. Test-baiting, like treatment, is only effective if moist, mild conditions prevail. When damage is expected to be heavy do not wait for the results of test-baiting but apply the treatment immediately. Further details are given in ADAS Leaflet 115, *Slugs and snails*.

Cereal Growth Regulators

The main crop protection aspect of plant growth regulators in winter cereals is the control of lodging by reduction of straw length and straw strengthening. Yield responses to growth regulators have been obtained from time to time in the absence of lodging and some product recommendations are made to maximise the chance of this rather than for lodging control. The recommendations summarised in Table 6.12 are those for straw shortening and strengthening only. The use of 2-chloroethylphosphonic acid and the mixture of that with mepiquat chloride for lodging control is only recommended on crops at risk such as those grown under high input systems, in situations of high fertility or when weak strawed varieties are grown.

Where chlormequat is used at early growth stages of wheat or barley to manipulate the pattern of growth, the other two chemicals may be applied for lodging control as recommended. On durum wheat it is suggested that chlormequat is applied as a split treatment where the lodging risk is high.

TABLE 6.12 Timing recommendations for lodging control chemicals in winter cereals

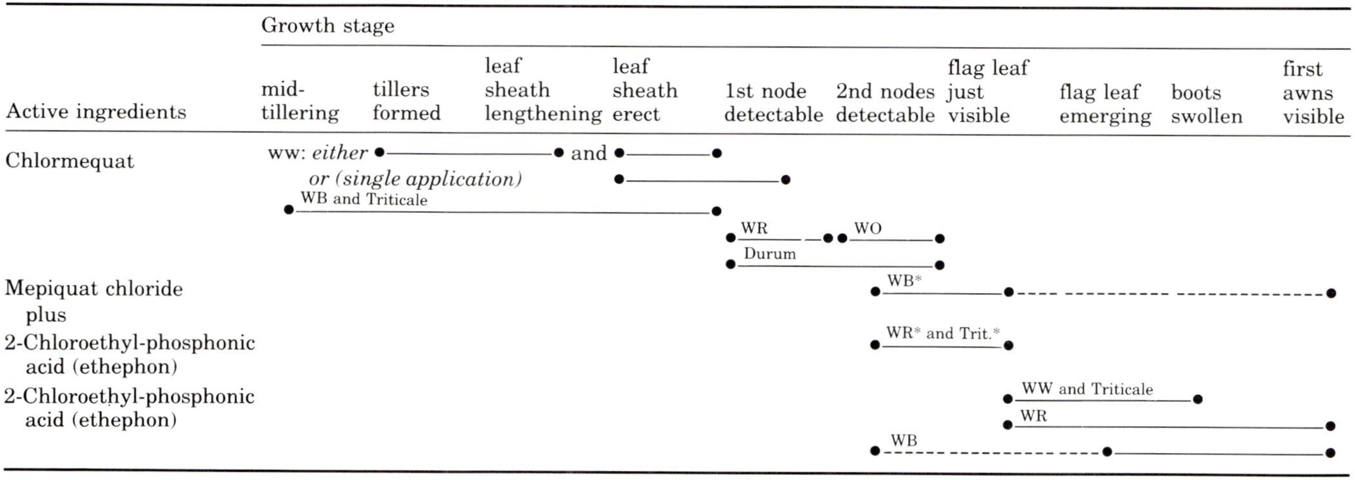

*Preferred timing after earlier application of chlormequat is at flag leaf just visible stage.

WW winter wheat
WB winter barley
WR winter rye
WO winter oat
Trit. triticale

TABLE 6.13 Guide to crop protection planning and crop scouting: autumn-sown cereals

Weeds and growth regulation	Diseases	Pests
1. Between crops and sowing period		
Cultivations for control of rhizomatous weeds; also herbicides (p. 115). Consider effect of primary cultivation in relation to dominant weed type (p. 7 and p. 113). Consider: effect of straw disposal method on weed seeds and soil acting herbicide use (p. 116); check variety tolerance against desired herbicide use; effect of cloddiness of seedbed and soil moisture status on soil acting herbicide effectiveness (p. 76); effect of sowing date on weed problems and residual effect expected from soil-acting herbicide use (p. 52 and p. 76). Residual herbicide application possible (p. 118).	Consider: diversification of cereal varieties and disease resistance of varieties selected (p. 132); control of volunteer cereals to reduce 'green bridge' for disease (p. 131); seed treatments available against seed-borne diseases (p. 134); consider especially treatment of home-saved seed. Winter barley – consider use of seed treatment against mildew especially in early sown high risk situations (p. 135).	Control volunteer cereals and grass weeds to reduce opportunities for egg-laying pests (p. 156). Consider effect of early stubble cultivations providing opportunity for wheat bulb fly egg laying (p. 160). Fineness of tilth and seedbed consolidation to reduce slug and wireworm (p. 156). Seed treatments available against wheat bulb fly (winter sown crops only) also against wireworm and frit fly (p. 157). Bait application against leatherjacket and slug prior to sowing (p. 158). In land known to be infested with stem nematodes avoid rye and non-resistant oat varieties (p. 41). Pest forecasts available from Advisory Services.
2. Emergence to tillering period (winter)		
Inspect fields for weed emergence Early post-emergence sprays possible – generally most effective time for annual grass weed control especially black-grass (p. 119). Consider spraying high population of broad-leaved weeds and any wild-oat (if previous populations have been high) (p. 117).	Inspect for foliar disease at regular intervals until growth stops in winter. Check for barley yellow mosaic symptoms in late winter and early spring. *Winter Barley:* control serious leaf-diseases present in September to December. Opportunities for spraying mildew, rhynchosporium and net blotch (p. 136) also to check spread in January to March period if	Inspect regularly for pest damage. Crops thinned by soil pests – consider rolling and top dressing when season suitable. Spray treatments available against frit fly, wheat bulb fly and yellow cereal fly (p. 159). If yellow cereal fly damage anticipated use spray at or before egg hatch (p. 162).

	youngest growth affected. *Winter wheat:* not normally economic to spray foliar disease at this time unless very severe. Check for eyespot in January/February. *Winter oats:* inspect for leaf diseases. *All cereals:* barley yellow dwarf virus: spray early sown crops in high risk areas with aphicide (p. 155 and p. 161).	Aphid found – consider barley yellow dwarf virus risk (p. 161). Pest forecasts available from Advisory Services for most areas.
3. Tillering to jointing period (spring)		
Inspect regularly to determine growth stage especially early drilled crops which may pass safe stage for spraying many herbicides (Table 6.4 and p. 117) before normal spraying season. Check effectiveness of autumn applied herbicides. Check weed emergence. Opportunities for use of wide range of herbicides: select according to weeds' present density, crop safety (species and variety tolerance, and growth stage) (p. 117 and tables following). *Winter wheat:* critical stage, becoming safe to spray with many herbicides is when main shoot leaf sheaths measure 5 cm from ground to base of last unfolded leaf. *Lodging control:* opportunity to spray crops at risk with chlormequat (all winter-sown cereals) see Table 6.12.	*Winter wheat:* Inspect for eyespot particularly between leaf sheath erect and first joint stage spray if necessary (p. 138). Inspect for foliar disease. If mildew above 10% threshold, spray (p. 140). If septoria lesions obvious use spray if not already sprayed against eyespot. Yellow rust: spray susceptible varieties when disease first seen. *Winter barley:* inspect for eyespot between leaf sheath erect and first node stage. Spray if threshold exceeded (p. 138). At stem erect stage inspect for foliar disease and spray if necessary (p. 139). *Winter oats:* inspect for crown rust (p. 153) and mildew (p. 139). *Rye:* inspect for eyespot between start of spring growth and first node stage; spray if necessary (p. 139).	Crops suffering pest attacks will be helped by nitrogen top dressing and rolling where possible. Opportunity for spraying to combat tiller damage by wheat bulb fly larvae and spray or bait control of leatherjackets. Check for aphid build-up on cereal leaves.
4. Stem elongation to crop maturation period		
Fewer opportunities for spraying broad-leaved weeds (p. 123) but materials available for spraying wild oat up to 3 nodes of barley or	*Winter wheat:* maintain rainfall records to determine 'wet periods' (p. 143). At 2 nodes to flag leaf visible stage check especially for	Check for presence of cereal aphids. Spray if threshold numbers exceeded (p. 162). Consider treatment of wheat blossom

TABLE 6.13—Continued

Weeds and growth regulation	Diseases	Pests
4 nodes wheat (p. 124). Rogue low wild-oat population (p. 126). Note dormant weed population and distribution for action in following years. Opportunity for spraying couch grasses (and other weeds) with glyphosate in wheat, barley and oats (p. 128), except those for seed or destined for distiller's malt. *Lodging control:* opportunities for spraying crops at risk with 2-chloroethylphosphonic acid and its mixture with mepiquat chloride. See Table 6.12.	mildew and yellow rust giving priority to susceptible varieties. Spray if necessary. Use broad-spectrum material to include septoria control if there has been a wet period (p. 143). When flag leaf emerged – critical period for development of severe leaf diseases starts. Inspect for all leaf diseases particularly septoria. Spray if necessary (p. 143). When ears just emerged consider need for 'insurance' spray if broad-spectrum fungicide not applied earlier (p. 143). Ears fully emerged to grain water ripe – check for and control late leaf and ear diseases especially rusts and septoria (p. 143). Seed crops: inspect for loose smut in crops and surrounding wheat crops (p. 134). At flag leaf to half ear emerged (GS 39–55); inspect regularly and spray to prevent spread of late leaf disease if necessary. From ear fully emerged to end of flowering (GS 58–69): inspect particularly for net blotch and brown rust. *Seed crops:* inspect at ear emergence for loose smut in crop and surrounding barley crops. *Winter oats:* crown rust spraying opportunity up to time of panicle emergence (p. 142); mildew sprays can be applied up to the boot swollen stage (p. 142). *Rye:* inspect for leaf disease between start of spring growth and flag leaf emergence; spray if necessary (p. 142).	midge, thrips or saddle gall midge found in large numbers (p. 160).

Chapter 7 CROP PROTECTION OPPORTUNITIES IN SPRING-SOWN CEREALS

The comments at the start of Chapter 6 also apply to this Chapter. However, the growth stages of spring-sown cereals being more condensed than those with winter-sown cereals, it is not convenient to divide this Chapter into the same broad growth stage periods in each of the sections dealing with weeds, diseases and pests.

Weed Control

In spring cereals weeds germinate more uniformly and their control is less complex than in winter cereals. A spring cereal can provide a weed control break for weed species of winter cereals which naturally germinate in the autumn or winter. To achieve this avoid early spring drilling in low temperatures when there is a risk of weeds such as black-grass, brome grasses and cleavers germinating in the spring cereals, giving rise to situations more akin to winter cereals but for which there are fewer control opportunities.

Cultural opportunities: autumn, winter and sowing period

Encourage natural losses of weed seeds from the previous crop by leaving stubbles undisturbed during the autumn months. Volunteer cereals or perennial grasses may be destroyed by spraying stubbles with paraquat with or without diquat, or glyphosate. Delay ploughing until December to give the maximum loss of seeds. Late ploughing will also avoid the establishment of large over-wintering weeds which are more difficult to kill before sowing the spring cereal.

Shallow cultivations or direct drilling rather than ploughing will have a similar influence on weed populations as described for winter cereals (see p. 113), except that annual grass weeds build up less rapidly than in winter cereals if stubbles are left uncultivated beforehand to enhance natural seed losses

Avoid transplanting weeds, especially grasses such as rough meadow-grass, through the use of inadequate cultivations. Under dry conditions seedbed cultivations alone may be sufficient to kill weeds but in wet seasons the use of herbicides as suggested below may be necessary.

As an opportunist measure it may be possible, in some years, to use a 'stale seedbed' technique to control the first flush of weeds. This involves preparing a seedbed as far as possible some time before sowing and killing weeds that grow with a final cultivation just before sowing or, if cultivation is unsuitable by spraying with a contact herbicide. It is normally more cost-effective to sow spring cereals as soon as soil conditions permit and to rely on herbicides for weed control than to use this technique and suffer the yield penalty accruing from late sowing.

Herbicides

Pre-sowing

Where seed bed cultivations alone are not sufficient to kill over-wintered or early germinating weeds, for instance in a wet spring, use a contact herbicide such as glyphosate, or paraquat with or without diquat to control them.

Pre-emergence

Methabenzthiazuron. This chemical can be used, in spring barley only, to control meadow-grasses and a limited range of broad-leaved weeds but including red dead-nettle

Tri-allate. The liquid formulation can be applied to spring barley as a soil-incorporated treatment both pre- and post-drilling to control wild-oats, meadow-grasses and moderate populations of black-grass. The barley needs to be drilled under at least 4 cm of settled soil. The pre-drilling treatment is preferred for wild-oat control.

The granular formulation of tri-allate can be used for the same weed range in both spring wheat and spring barley. For spring wheat apply the granules immediately after drilling and any subsequent harrowing or rolling and do not incorporate. Drill the wheat to give at least 4 cm of settled soil cover. For spring barley either pre- or post-drilling application can be made but pre-drilling distribution with subsequent incorporation by drilling and harrowing gives the best results. Barley is more tolerant than wheat and the cover of settled soil can be between 2.5 and 4 cm.

Linuron. This is recommended for pre-emergence use in spring-sown crops of wheat, barley and oats, principally for corn marigold control.

Pendimethalin. This is recommended for pre-emergence use in spring barley for annual meadow-grass and a wide range of broad-leaved weeds control.

Soil moisture. All these treatments require adequate moisture in the seedbed to maximise activity, particularly the non-incorporated granule treatment.

Post-emergence

A wide range of herbicides is available for use in spring cereals (Tables 7.1 and 7.2). All can be used in spring barley; all except bifenox/mecocrop in spring wheat, and the majority in spring oats. Select on the basis of safety to the crop (taking into account the particular growth stage), and the weeds present. Care must be taken not to apply hormone herbicides too early, particularly MCPA, 2,4-D, dicamba and 2,3,6-TBA. Application during ear initiation can result in ear deformities which may result in grain loss. Although weed species may be anticipated from previous years, fields should be walked to confirm the species and numbers of weeds after the main flush has emerged. Unexpected weed populations may arise especially through the exposure of long-buried seed, such as from poppies, by deeper than normal soil disturbance and through the effects of different weather patterns from season to season. Guidance on the concept of threshold levels on which to base spray decisions is given in Chapter 2 (p. 10). Choose the herbicide best able to control the main and most competitive weeds present.

Guidance on herbicide selection, with comparative weed lists, is given in the publications listed on p. 109. More detailed weed lists are obtainable from commercial literature for individual products.

Of the herbicides recommended for post-emergence broad-leaved weed control those containing bifenox, cyanazine or linuron have a limited amount of residual activity which may extend weed control activity for a few days after spraying; those containing metsulfuron-methyl or chlorsulfuron have a much longer effective residual life. This confers extended weed control activity but also requires precautions to be taken as regards succeeding crops, particularly in the event of a crop failure. The recommendations on the label should be studied and followed carefully.

Late spraying

Where spraying is unavoidably delayed past the jointing stage additional care is needed in selecting a herbicide for broad-leaved weed control that is safe to the crop. Materials that are recommended for spraying after the jointing stage are shown in Table 7.1.

For late wild-oat control benzoylprop-ethyl can be applied to spring wheat, and flamprop-M-isopropyl to spring wheat and spring barley, both up to but before the fourth node detectable stage. Other wild-oat herbicides can be applied at late crop stages (see table 6.6) but are generally less effective on large wild-oats.

TABLE 7.1 Herbicides for broad-leaved weed control in spring cereals (including some recommended principally for grass weed control)

Active ingredient	*Use and earliest timing in spring-sown crops of:* wheat	barley	oats	*Recommended for use after jointing*	*Use on crops undersown with:* grass	clover	lucerne
benazolin + 2,4-DB + MCPA	5 lvs	2 lvs	2 lvs	no	√	√	no
bentazone + dichlorprop	3 lvs	3 lvs	3 lvs	yes up to 1 node	√	no	no
bentazone + dichlorprop + MCPA	4 lvs	4 lvs	4 lvs	yes up to 2 nodes	√	√	no
bentazone + MCPB + MCPA	2 lvs	2 lvs	2 lvs	yes up to 1 node	√	√[1]	no
bifenox + mecoprop	no	3 lvs	no	yes up to 1 node exceptionally 2 nodes	no	no	no
bromoxynil + ioxynil	1 leaf	1 leaf	1 leaf	yes up to 2 nodes	√	√[2]	no
bromoxynil + ioxynil + benazolin	2 lvs	2 lvs	2 lvs	yes up to 2 nodes	√	no	no
bromoxynil + ioxynil + dichlorprop	3 lvs	3 lvs	3 lvs	some products to 1 node	no	no	no
bromoxynil + ioxynil + mecoprop esters	1 leaf	1 leaf	1 leaf	yes up to 1 node exceptionally 2 nodes	rye-grass only	no	no
bromoxynil + ioxynil (esters) + mecoprop salt	3 lvs	3 lvs	no	yes up to 1 node	rye-grass only	no	no
chlorsulfuron + ioxynil + bromoxynil	2 lvs	2 lvs	no	yes up to 2 nodes	no	no	no
clopyralid + bromoxynil + MCPA *or* + dichlorprop + MCPA	5 lvs	5 lvs	5 lvs	no	no	no	no
clopyralid + mecoprop ± bromoxynil	1 leaf	1 leaf	1 leaf	yes up to 1 node exceptionally 2 nodes	no	no	no
clopyralid + ioxynil + mecoprop ± bromoxynil	1 leaf	1 leaf	1 leaf	with bromoxynil: no otherwise to 2 nodes	no	no	no
cyanazine + MCPA	4 lvs	4 lvs	2 lvs	yes up to 1 node	no	no	no
cyanazine + mecoprop	4 lvs	4 lvs	2 lvs	yes up to 1 node	no	no	no
cyanazine + clopyralid	√	√	√	tank-mix: timing depends on products added	no	no	no
2,4-D	5 lvs	5 lvs	no	no	√	no	no
2,4-DB	3 lvs	3 lvs	3 lvs	no	√	√	√
2,4-DB + linuron + MCPA	5 lvs	2 lvs	2 lvs	no	√	√	no
2,4-DB + MCPA	4 lvs	2 lvs	2 lvs	no	√	√	no
2,4-DB + mecoprop	1 tiller	2 lvs	2 lvs	yes up to 1 node	√	√	no

dicamba + dichlorprop + MCPA	5 lvs	5 lvs	5 lvs	no	√	no	no
dicamba + mecoprop	5 lvs	5 lvs	5 lvs	no	√	no	no
dicamba + mecoprop + MCPA	5 lvs	5 lvs	5 lvs	no	√	no	no
dichlorprop	1 leaf	1 leaf	1 leaf	no	no	no	no
dichlorprop + bromoxynil	2 lvs	2 lvs	2 lvs	yes up to 1 node	no	no	no
dichlorprop + 2,4-D	5 lvs	5 lvs	no	no	no	no	no
dichlorprop + MCPA	5 lvs	3 lvs	1 leaf	no	no	no	no
dichlorprop + MCPA + ioxynil	3 lvs	3 lvs	3 lvs	no	no	no	no
dichlorprop + mecoprop	1 leaf	1 leaf	1 leaf	no	no	no	no
dinoseb amine, acetate	3 lvs	3 lvs	3 lvs	no	√	√	√
fluroxypyr	2 lvs	2 lvs	2 lvs	yes up to 1 node	√	no	no
isoproturon + bromoxynil + ioxynil*	2 lvs	2 lvs	no	yes up to 1 node	no	no	no
isoproturon + ioxynil + mecoprop*	2 lvs	2 lvs	no	yes up to 1 node	no	no	no
ioxynil + mecoprop (esters)	2 lvs	2 lvs	3 lvs	oats: no	no	no	no
				wheat and barley to 1 node	no	no	no
linuron	pre-em only	pre-em only	pre-em only	—	no	no	no
MCPA	5* lvs	5* lvs	5* lvs	no	√	(√)[3]	no
	*half-dose from 3 leaf stage						
MCPB	1 leaf	1 leaf	1 leaf	no	√	some	no
MCPB + MCPA	2 lvs	2 lvs	2 lvs	no	√	√	no
mecoprop	1 leaf	1 leaf	1 leaf	yes up to 1 node	√	no	no
mecoprop + ioxynil + benazolin	2 lvs	2 lvs	2 lvs	yes up to 2 nodes	√	no	no
methabenzthiazuron*	no	pre-em only	no	—	rye-grass only	no	no
metsulfuron-methyl	2 lvs	3 lvs	no	yes up to 2 nodes	no	no	no
pendimethalin*	no	post-drilling	no	no: latest time early tillering	no	no	no
2,3,6-TBA + dichlorprop + MCPA	3 lvs	3 lvs	3 lvs	no	no	no	no
2,3,6-TBA + dicamba + mecoprop + MCPA	5 lvs	5 lvs	5 lvs	no	√	no	no

*some grass weed control included
Note: This Table contains major product mixes but does not include all the herbicide mixtures found on product labels.

1. grass/clover mixes only.
2. when used pre-sowing or pre-emergence of the clover.
3. grass/clover mixes only – very variable product recommendations.

TABLE 7.2 Herbicides for wild-oat control in spring cereals

Active ingredient	Use in spring-sown crops of: wheat	barley	Undersown situations:	Effects on other grass weeds
Pre-emergence				
tri-allate *liquid*	no	√	safe to clover only	Controls meadow-grasses and suppresses a number of other grass species, including black-grass.
granular	√	√	safe to clover only	
Post-emergence				
benzoylprop-ethyl	√	no	yes[1]	
diclofop-methyl[2]	√	√	no	Controls rye-grass and rough meadow-grass but not annual meadow-grass. Suppression of black-grass.
difenzoquat	√	√	yes[1]	
flamprop-methyl	√	no	yes[1]	Suppresses meadow-grasses and black-grass.
flamprop-M-isopropyl	√	√	yes[1]	Suppresses onion couch and black-grass.

1. Can be used on crops undersown with: Italian rye-grass, perennial rye-grass, ryegrass/clover and clover.
2. Not recommended in northern England and Scotland in the spring.

Undersown Cereals

Undersowing grass seeds in cereals limits the choice of herbicides which may be used subsequently, especially if clover is included in the mixture. Undersowing with clover alone or with lucerne limits choice even further (see Table 7.1).

The germination of seeds sown into the cereals will also be affected by the residual activity of herbicides applied earlier. Soil-acting herbicides which can be expected to affect subsequent undersowing possibilities are shown at Table 4.3 (p. 72). In addition some foliage-applied hormone herbicides such as mecoprop have sufficient residual activity in the soil to affect germinating seeds even though they are safe on seedling grasses. To overcome this risk it is usually recommended that in situations involving undersowing, herbicides (other than those such as bromoxynil and ioxynil with purely contact action) are used post-emergence of the undersown species when both they and the cereal are at safe growth stages. In some cases lower doses are recommended for undersown cereals; even so grass may suffer a temporary check if sprayed at the two–three leaf stage which does not occur when spraying is delayed until the early tillering stage. The latter is generally considered the safe stage for grasses but recommendations are made for spraying earlier. The safe stages for clovers and other legumes vary considerably. A brief guide is given in the Farmers Weekly Supplement, *Cereals Spring Herbicides*. Further guidance can be found in ADAS Booklet 2056, *Weed control in grassland, herbage legumes and grass seeds 1985-6*, (MAFF Pulications, Alnwick) and the *BCPC Crop Protection Handbook Grass and Clover Swards*, 1984, Ed. R. D. Williams, (BCPC Publications, Croydon).

Where cereals are to be undersown it is recommended that the cultural measures discussed earlier in this chapter and in Chapter 6 are taken before sowing the crop, to reduce the weed burden, especially when meadow-grasses and perennial weeds such as couch grass are present in the field.

Herbicides for undersown crops

Tri-allate. This is recommended for pre-emergence use in wheat and barley to be undersown with clover (not grass) mainly for wild-oat control as shown in Table 7.2. See also the paragraphs on p. 76 and 114 for further details.

Methabenzthiazuron. This can be used pre-emergence of spring barley and wheat that is undersown with rye-grass only, as soon as possible after drilling both the cover crop and the rye-grass. It is used primarily for the control of meadow-grasses and some broad-leaved weeds.

Bromoxynil and ioxynil. This combination of contact-acting herbicides can be sprayed either pre- or post-emergence of the sown grass species and pre-emergence only of sown clovers, lucerne, sanfoin and trefoil.

Post-emergence herbicides. The range available against broad-leaved weeds is indicated in Table 7.1. Those that can be used primarily for wild-oat, and self-sown oat control are shown at Table 7.2.

Information about controlling specific broad-leaved weeds can be found in the literature mentioned previously (see p. 109) and in product labels and guides.

Notes on Some Specific Weeds

Wild-oats

Providing over-wintered and early spring germinating wild-oats are killed before sowing wheat or barley in the spring they are likely to germinate more evenly than in winter crops and thus be somewhat easier to control. See p. 126 for further comments and reference. See Table 7.2 for wild-oat herbicide recommendations for use in spring sown wheat and barley.

Black-grass

Black-grass is not commonly a serious problem in spring-sown cereals but can be so if very early drilling is practised. Since there are only limited effective herbicide recommendations for controlling black-grass in spring cereals cultural measures are important where this problem occurs.

Avoid sowing too early and ensure that over-wintered plants are not transplanted in the seedbed by inadequate cultivations. See p. 126 for further comments and reference. Herbicide recommendations for black-grass control in spring wheat and barley include: isoproturon + ioxynil + mecoprop and isoproturon + bromoxynil used post-emergence; pendimethalin used pre-emergence (spring barley only).

Useful suppression of black-grass in spring wheat and spring barley is given by: tri-allate applied in the seedbed; diclofop-methyl post-emergence; flamprop-methyl post-emergence (not spring barley); and flamprop-M-isopropyl post-emergence. These materials are applied primarily for wild-oat control. See Table 6.6 (p. 124) for application timing.

Broad-leaved weeds

Herbicides which are especially useful for specific problem weeds are listed in an early section of MAFF Reference Book 380, *Approved Products for Farmers and Growers*, 1985 (HMSO, London).

Perennial weeds

Most of the herbicides which can be effective against perennial broad-leaved weeds are foliage-applied but are not suitable for use in the growing cereal crop. This is because the weeds seldom develop a large enough leaf area in relation to the underground storage organ before the crop grows past the safe stage for their use.

The use of glyphosate pre-harvest in crops not destined for seed or distillers' malt controls a wide range of perennial broad-leaved weeds as well as couch grasses. Glyphosate may also be used, less effectively, on undisturbed stubble. Other possible stubble treatments include the use of aminotriazole to foliage followed by thorough cultivation or the application of hormone herbicides such as clopyralid, 2,4-D, dicamba mixes, MCPA and 2,3,6-TBA mixtures.

Perennial grasses can be dealt with between crops as indicated on p. 115.

Disease Control

Sources of inoculum

The foliar pathogens which cause the main disease problems on spring-sown cereals may survive on debris or on infected volunteer plants. These sources of inoculum decline during the winter, however, and in areas where both spring and winter cereals are grown they are of less importance than is the carry-over of disease on autumn-sown crops. This is especially true of obligate parasites such as those causing mildew and the rusts. The development of these diseases is often about two weeks earlier in spring crops close to autumn-sown fields than in crops distant from such over-wintering sources.

It is advisable, therefore, not to situate spring-sown crops close to autumn-sown crops of the same species. If this is unavoidable the risks of early disease spread may be minimised by the application of fungicide to the winter crop before nearby spring crops emerge. This may lead to the winter crop being sprayed so early that a second application is necessary later in the spring.

Choice of variety

Disease factors to be considered in choosing spring cereal varieties are much the same as those discussed in the context of winter cereals (Chapter 6). It is particularly important to avoid varieties highly susceptible to any of the main foliar diseases if the crop is to be sown in a field adjacent to an autumn-sown crop of the same species. Spring barley varieties are assigned to Diversification Groups according to the genetic bases of their resistance to mildew (see NIAB Farmers leaflet No. 8, *Recommended varieties of cereals*). The growing of two or more varieties from different groups will reduce field to field spread of the disease.

If a “blend” of say, three varieties is grown, the level of mildew in the blend will be lower than the average level in crops of the three varieties grown separately. It should be noted that the level in the blend may not be so low as the level in the best of the three single variety crops. However, since it is impossible to predict which strain of mildew will occur on the farm, it is equally impossible to predict which of the varieties will have the lowest level of disease (assuming that the three varieties have comparable scores for disease resistance). It must be borne in mind that offering mixed varieties may limit marketing opportunities for grain.

Sowing date

Early-sown crops are more susceptible to both take-all and eyespot; but even the earliest of spring sowings suffer less from these diseases than do autumn-sown crops. Early sowing may also lead to early infection by foliar disease. However, the most severe attacks of mildew are often seen in “cuckoo barley” which is so late sown that it emerges when the disease is already well established on surrounding crops and so becomes heavily infected at the seedling stage.

Since they become infected at an earlier stage of growth late-sown spring cereals can also suffer more severely than earlier sowings from the effects of barley yellow dwarf virus (BYDV). Even in late-sown crops it is unusual to find more than scattered plants affected by the virus and it is seldom necessary to take protective action against the disease.

Spring varieties are susceptible to barley yellow mosaic virus (BaYMV) if they are sown in autumn but no variety, winter or spring, develops symptoms if it is spring-sown.

Seed treatment

Seed-borne diseases may be controlled by seed-applied fungicides as described for winter cereals (see p. 134 and Table 6.7). Seed treatments which will control mildew, or a wider range of foliar diseases, as well as seed-borne pathogens are now available for barley, oats and wheat. Their use on spring cereals may be particularly beneficial if a crop of a susceptible variety is sown in close proximity to an autumn-sown crop of the same species. If such a treatment is to be used it is advisable to use a chemical with a different mode of action from any fungicide used on the nearby winter cereal. Seed treatments available for the control of foliar diseases on spring barley are shown in Table 6.8 (p. 137); mode of action groupings are given in Table 4.7 (p. 88). Triadimenol + fuberidazole is recommended for use on spring wheat and spring oats as well as on barley (and winter wheat).

Foliage-applied fungicide treatments

The fungicide treatments available for spring cereals are shown in Table 6.9 (p. 140).

Spring barley: fungicide use

Late tillering to first node visible (GS 30/31)

If no seed treatment active against mildew was applied, then even if no disease is seen in the crop apply a mildew fungicide with any compatible herbicide if the crop is of a very susceptible variety and/or is very close to winter barley. Use a chemical with a different mode of action to any used on the winter barley.

Late tillering to flag leaf visible (GS 31-37)

Under the "high risk" conditions described in the preceding paragraph, the spray applied with the herbicide should be followed by a second application about three weeks later if fresh mildew pustules can be found on the plants or if rhynochosporium is seen on susceptible varieties (disease rating 4 or less).

Other crops should be inspected about every five days and an appropriate fungicide applied if mildew becomes obvious on any of the green leaves, or if rhynchosporium is seen on varieties with a disease rating of 4 or less.

Flag leaf emerged to ear emerged (GS 39-59)

Apply an appropriate fungicide or combination of fungicides if:

(i) mildew is present on any of the top three leaves;
(ii) if rusts, rhynchosporium or net blotch are present, even at low levels, in crops of susceptible varieties (disease rating 4 or less);
(iii) if rusts, rhynchosporium or net blotch are obvious on the top three leaves of more resistant varieties.

Start of flowering to end of flowering (GS 61/69)

Apply an appropriate fungicide *only* if brown rust, yellow rust or net blotch become obvious on the top three leaves of susceptible varieties. See Table 6.10 (p. 146) for latest recommended timings.

Fungicide programmes

In practice for all except the most susceptible varieties (and crops adjacent to winter barley) a single application of an effective fungicide when mildew first becomes obvious in the crop is often all that is necessary for spring barley. If two or more sprays are applied (e.g. where a seed treatment is followed by a spray) each application should be of a fungicide different in its mode of action from the material last applied. An alternative strategy is to use formulated mixtures (or recommended tank mixes) of fungicides with different modes of action (Table 4.7).

Spring wheat: fungicide use

Spring-sown crops are unlikely to suffer severely from eyespot and a spray at the late tillering stage is likely to be necessary only if mildew appears in the crop; this is most likely to occur where it is in close proximity to winter-sown crops of the same species.

Guidelines for the later application of fungicides to wheat are the same as those given for the autumn-sown crop in Chapter 6 from p. 138 onwards.

Spring oats

Avoid growing spring oats next to winter oats and control overwintered volunteer oat plants.

Crown rust: spray susceptible varieties (NIAB rating of 4 or less) if the disease is found on the top three leaves up to the time of panicle emergence; spray other varieties if the disease is spreading on these leaves. Fungicides available for control are fenpropimorph, triadimefon and triadimenol.

Oat mildew: spray at the stem extension stage if the disease is obvious on the lower leaves of most plants. (There may be the opportunity of tank mixing with herbicide.) Use the same criterion for later growth stages but do not spray after panicle emergence. Where an early spray does not contain the disease a second spray applied 3 or 4 weeks later may be justified. Fungicides available: fenpropimorph nuarimol, triadimephon and tridimorph.

If crown rust is present on the plants at the time of spraying mildew the preferred materials are fenpropimorph triadimenol and triadimefon. Seed treatment with triadimenol plus fuberidazole gives protection against mildew as well as all the important seed-borne diseases (see Table 6.7).

Pest Control

Much of the material given in the Pest Control section of Chapter 6 is relevant to spring-sown crops and should be read in conjunction with the paragraphs below.

General cultural measures

Destruction of egg-laying sites. Destruction of grass weeds and volunteer cereals in the autumn will greatly reduce the risk of damage to spring cereals by leatherjackets, frit fly, crambid and swift moth caterpillars and similar pests.

Seedbed preparation. A fine, firm seedbed reduces the damage caused by wireworm and slugs where these pests are present. Adequate crop nutrition also helps to overcome soil-pest damage.

Time of sowing. Obtain local advice from official advisory services on the likelihood of large-scale hatchings of pests such as frit fly and wheat bulb fly affecting cereals. Early sowing of spring oats is needed to counter frit fly attack. Spring wheat and spring barley should not be drilled too early. Further details are given in the section on specific pests below.

Where gout fly is a problem, early sowing of spring barley is important to avoid serious damage.

Recommendations for control of specific major pests

Wheat bulb fly (*Delia Coarctata*)

Avoid sowing spring wheat or spring barley before mid-March on land at risk (i.e. where previous crop was fallow, root crop or vining peas). Spring oats are immune to attack.

There are no satisfactory chemical treatments for use in spring cereals. Insecticidal seed treatment should not be used on spring-sown cereals.

Cereal aphids.

Except in some seasons in western parts of England and Wales, barley yellow dwarf virus disease is of little importance in spring-sown cereals. Aphicidal granules or sprays applied at drilling or shortly afterwards have occasionally given some reduction in virus spread but with little yield benefit.

The importance of cereal aphids infesting spring cereals lies in their direct feeding damage.
Some varieties of spring wheat support lower numbers of the grain aphid and/or rose-grain aphid and probably suffer less damage in consequence. Spring barley is more tolerant of aphid damage than spring wheat or spring oats.

It is recommended that crops are examined from flag leaf emergence onwards. If five or more grain aphids are present on the ear, or if 30 or more rose-grain aphids infest the flag leaf, and the weather is warm and settled, apply single sprays of one of the following aphicides, using preferentially those which have least deleterious effects on beneficial organisms such as bees (e.g. phosalone, pirimicarb): chlorpyrifos, demeton-S-methyl, dimethoate, formothion, heptenophos, oxydemeton-methyl, phosalone, pirimicarb, thiometon. Some of these insecticides may be applied at a lower dose for grain aphid control after ear emergence and before flowering. Treatments applied after the milky/cheesy ripe stage are of little value.

Frit fly (*Oscinella frit*)

Early sowing (mid-March in southern England to late March in the north and Wales) of oats is most effective; however, early-sown oats which are backward in growth may be attacked.

For late-sown oats, or oats which have not reached the four-leaf stage by early to mid-May (southern England) or late May to early June (north and Wales), a spray application of chlorpyrifos or triazophos may be necessary. Local advisory officers will give advice on the likelihood of frit fly damage.

Leatherjackets (*Tipula* spp. larvae)

Spring cereals following grass or grassy stubbles are at risk in years when leatherjacket numbers are high; consult official advisory service for forecast. For cereals following grass, make sure that the grass is ploughed and well buried before egg laying starts in September. Sow into a well consolidated seedbed with a good tilth. Roll and top-dress with nitrogenous fertiliser to help stricken crops to recover.

If large numbers of leatherjackets are known to be present, apply a spray of gamma-HCH or chlorpyrifos to the soil and lightly incorporate the insecticide before sowing.

Granules of chlorpyrifos can be applied as a broadcast treatment when damage is first seen or when sampling reveals 15 or more leatherjackets in 10 separate 30 cm lengths of drill.

Poison baits containing fenitrothion or gamma-HCH be prepared or purchased ready-mixed and these are broadcast over the soil surface to give effective leatherjacket control. Methiocarb pellets, applied primarily for slug control, also deal with leatherjackets.

Recommendations for sprays of chlorpyrifos, etrimfos, gamma-HCH, quinalphos or triazophos are available but these are not as effective as poison baits. They should be applied when the soil is damp and temperatures are above 6°C.

Gamma-HCH baits and sprays should not be used where potatoes or carrots are to follow within 18 months, because of the risk of taint.

Further details are given in ADAS leaflet 179.

Wireworms (*Agriotes* spp. larvae)

Oats and wheat are more susceptible than barley or rye. Attacks are worse in the first three or four years after ploughing old pastures, and are accentuated in poor seedbeds which are acidic in reaction or which carry crop diseases. Sow into a firm, fine seedbed following thorough cultivations, and supply adequate nutrients for rapid growth. Attacked crops should be rolled and top-dressed with nitrogenous fertiliser.

Gamma-HCH spray applied to the soil during seedbed preparation will deal with large populations of wireworms. Do not plant potatoes or carrots within 18 months of such an application. Seed treated with dry powder or liquid formulations of gamma-HCH can be used for small to moderate wireworm numbers. Poor quality or damaged seeds, or samples of seed containing more than 16% moisture, should not be treated as injury to the seed can occur. Sow treated seed as soon as possible. Fonofos, used primarily against wheat bulb fly, gives incidental protection against wireworms. Further details are available in ADAS Leaflet 199, *Wireworms*.

Cereal cyst nematode (*Heterodera avenae*)

Oats are most susceptible, and rye is least susceptible, to damage by this nematode. Barley is more tolerant than wheat and is preferred for cropping in infested fields; varieties of spring barley are now available which are resistant to one or more races of the pest (consult ADAS or NIAB Recommended List). Numbers of the nematode (consult ADAS for advice on soil sampling) fall dramatically under grass and a break of three or more years can provide an effective means of reducing infestations. Although resistant oat varieties (winter and spring) are known, they are sensitive to nematode invasion and do not yield well in infested soils.

A number of nematicidal materials have been tested successfully against cereal cyst nematode but they are not cost-effective at current prices. For further details see ADAS Leaflet 421, *Cereal cyst nematode*.

Stem nematode (*Ditylenchus dipsaci*)

Wheat and barley are not attacked in Great Britain and can safely be grown on infested land. Many oats and rye varieties suffer severe damage but a few oat varieties are known to be resistant. As the pest can survive in many crops and weed plants, it is virtually impossible to eradicate by cultural means. Provide adequate nutrients especially potash in the seedbed to assist partial recovery from attack. No chemical control measures are recommended.

TABLE 7.3 Guide to crop protection planning and crop scouting: spring-sown cereals

Weeds	Diseases	Pests
All cereals: Monitor growth stage and emergence of weeds to assist in selection of appropriate herbicide where use is necessary.	Seed treatments available against seed disease and mildew (p. 178).	When leatherjacket counts high consider early ploughing grass that precedes spring cereals (p. 182).
Critical stage for starting herbicide spraying of spring cereals is assessed by leaf number. Important time for monitoring weeds is in the crop three leaf to five leaf stage (Table 7.1).	*Spring Barley:* check for early mildew especially crops near winter barley. Spray if necessary (p. 179).	Sprays against leatherjackets available.
Wild-oats: consider hand-pulling very low populations (p. 126).	Opportunity for combining herbicide and mildew spray up to GS 32.	Well consolidated seed-beds and good tilth help overcome soil pests.
Note weeds present when assessing disease at GS 39–59 for future reference	From 2 node to flag leaf (GS 32–37) – continue checking for mildew and spray if necessary. Check for rhynchosporium especially on susceptible varieties and spray if necessary.	Where wheat bulb fly risks high avoid sowing spring barley and spring wheat before mid-March; no spray treatments available; spring oats immune (p. 181).
Couch grass and other weeds in maturing crop: opportunity for applying glyphosate pre-harvest (p. 128) in crops not destined for seed or distillers' malt.	From flag leaf emerged to ear emerged (GS 39–59). Check mildew on top three leaves: check rusts and net blotch presence on plant and note which leaves affected: spray if necessary (p. 179).	Early sowing spring oats helps overcome attack by frit fly (p. 182).
	At flowering period (GS 61–69) check for late attack on net blotch, yellow and brown rust on top 3 leaves. Spray susceptible varieties if	Consider cereal species change if wireworm present in land ploughed from old grass (p. 183). Chemical spray available.
		Cereal species also differ in susceptibility to cereal cyst nematode and resistant varieties of spring barley available.

present (p. 179). Check seed crops and surrounding barley for loose smut (p. 135). *Spring wheat:* Check for mildew at early tillering especially crops next to winter wheat. Inspect for all leaf diseases at flag leaf emergence. Spray if necessary. Consider need for 'insurance' spray at ear-emergence if broad-spectrum material not applied earlier. When ears fully emerged: check for rusts and septoria and other late diseases on top leaves; seed crops – inspect for loose smut in crop itself and surrounding wheat crops. *Spring oats:* inspect for all leaf diseases from early tillering onwards. Spray crown rust on susceptible varieties as soon as seen. Spray against mildew if the disease is found to be spreading to the top 3 leaves.	Stem nematode: where land known to be infested avoid non-resistant oat varieties (p. 183). Cereal aphids: examine crops from flag leaf emergence onwards and spray serious infestations especially on spring barley (p. 181).

Cereal Growth Regulators

Growth regulators can afford protection to spring cereals where they reduce lodging by shortening and strengthening straw. Recommendations are made for the use of chlormequat for this purpose in spring wheat and spring oats.

The mixture of 2-chloroethylphosphonic acid (ethephon) with mepiquat chloride is recommended for use in specified barley varieties in Scotland and Northern Ireland only.

Appendix I

Diagrams of Important Growth Stages of Cereals

	Seedling growth			**Tillering**		
	First leaf through coleoptile	1st leaf unfolded	2 leaves unfolded	3 leaves unfolded, main shoot and 1 tiller	5 leaves unfolded, main shoot and 3 tillers	Leaf sheath (pseudostem) erect
Decimal key code	10	11	12	13, 21	15, 23	30

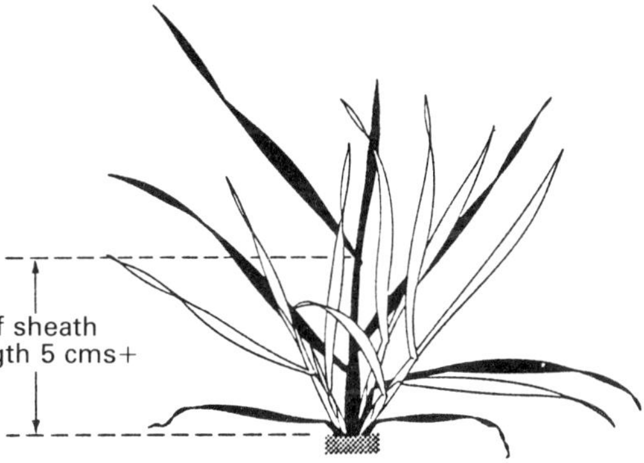

Measurement to define pseudo stem erect in winter wheat only

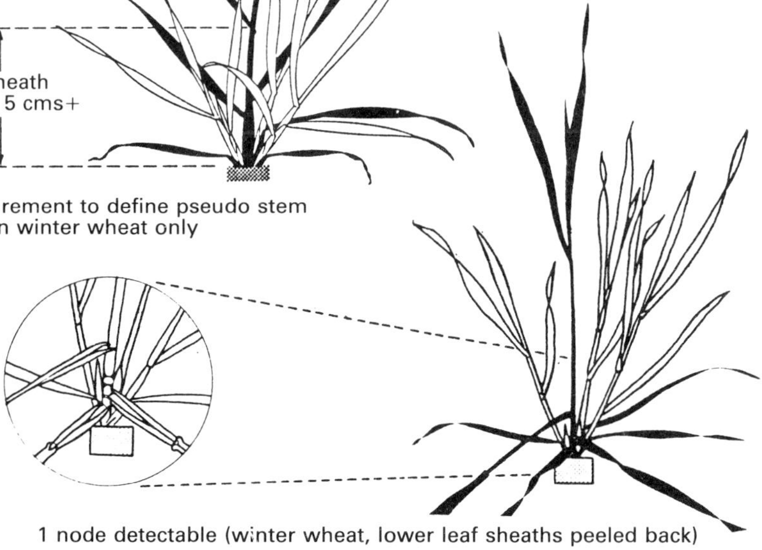

1 node detectable (winter wheat, lower leaf sheaths peeled back)

Illustrations by H. R. Broad, Weed Research Organisation and E. C. Large

In the stylised drawings below the tillers have been omitted

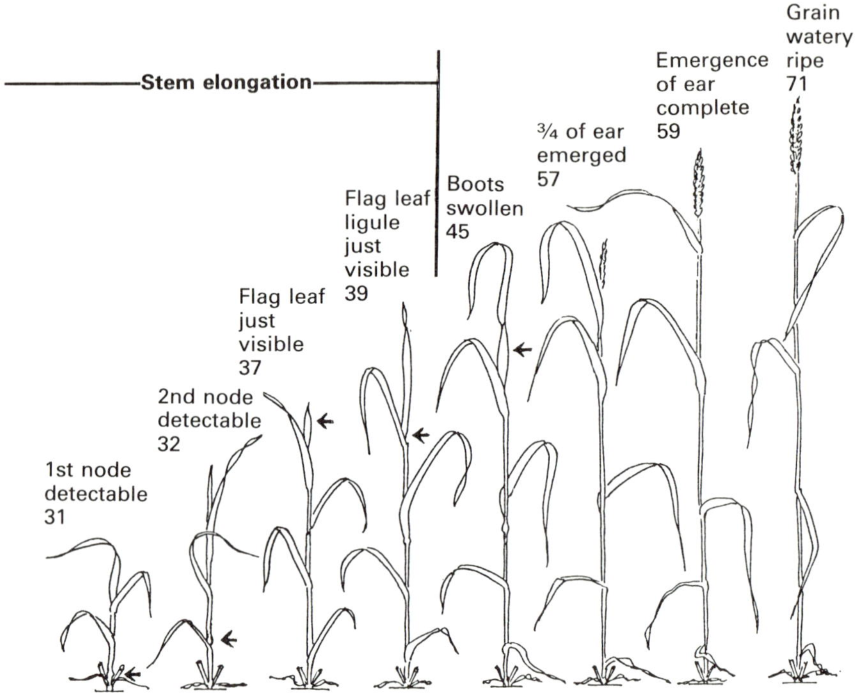

For more detailed explanation please refer to *An Explanation of the Decimal Code for the Growth Stages of Cereals, with Illustrations*, BCPC Occasional Publication No. 2.

Appendix II

The Decimal Code for Cereal Growth Stages

Definitions for principal and secondary growth stages

0 *Germination*

1 *Seedling growth*
- 10 First leaf through coleoptile
- 11 First leaf unfolded
- 12 2 leaves unfolded
- 13 3 leaves unfolded
- 14 4 leaves unfolded
- 15 5 leaves unfolded
- 16 6 leaves unfolded
- 17 7 leaves unfolded
- 18 8 leaves unfolded
- 19 9 or more leaves unfolded

2 *Tillering*
- 20 Main shoot only
- 21 Main shoot and 1 tiller
- 22 Main shoot and 2 tillers
- 23 Main shoot and 3 tillers
- 24 Main shoot and 4 tillers
- 25 Main shoot and 5 tillers
- 26 Main shoot and 6 tillers
- 27 Main shoot and 7 tillers
- 28 Main shoot and 8 tillers
- 29 Main shoot and 9 or more tillers

3 *Stem elongation*
- 30 Pseudostem (leaf sheath) erection
- 31 First node detectable
- 32 2nd node detectable
- 33 3rd node detectable
- 34 4th node detectable
- 35 5th node detectable
- 36 6th node detectable
- 37 Flag leaf just visible
- 39 Flag leaf ligule just visible

4 *Booting*
- 41 Flag leaf sheath extending
- 43 Boots just visibly swollen
- 45 Boots swollen
- 47 Flag leaf sheath opening
- 49 First awns visible

5 *Inflorescence (ear/panicle) emergence*
- 51 First spikelet of inflorescence just visible
- 53 ¼ of inflorescence emerged
- 55 ½ of inflorescence emerged
- 57 ¾ of inflorescence emerged
- 59 Emergence of inflorescence completed

6 *Anthesis (flowering)*
- 61 Beginning of anthesis
- 65 Anthesis half-way
- 69 Anthesis complete

7 *Milk development*
- 71 Caryopsis (kernel) water ripe
- 73 Early milk
- 75 Medium milk
- 77 Late milk

8 *Dough development*
- 83 Early dough
- 85 Soft dough
- 87 Hard dough

9 *Ripening*
- 91 Caryopsis hard (difficult to divide)
- 92 Caryopsis hard (not dented by thumbnail
- 93 Caryopsis loosening in daytime
- 94 Over-ripe, straw dead and collapsing

Note.
Growth stages 31 and 32 are critical limits for many herbicide applications. They have been commonly distinguished by feeling for nodes. It is proposed that a more objective measurement should be made by splitting the stem and measuring the distance between nodes. GS31 is then defined as starting when the distance between the crown and first node exceeds 2 cms and GS32 when the distance between the first and second nodes exceeds 2 cm.

At the time of writing the majority of organisations and companies agree to these proposed definitions but there remain a number of product recommendations still to be checked against the new definition.

Appendix III

List of Proprietary Names

This appendix is provided to help in the translation of active ingredients into product names. Although efforts have been made to include as many names as possible it is not exhaustive. The omission of names is not meant to imply any criticism of the products concerned. Where there are many suppliers of a particular chemical the various brands are not listed since readers should have no difficulty in locating them; in many of these cases the chemical name forms part of the product name.

A regularly updated list of products cleared for agricultural and other uses in the UK by the Pesticide Safety Precautions Scheme is available from the Ministry of Agriculture Fisheries and Food, London. This, in an expanded form, is expected to provide the basis of a new MAFF publication as a result of the new legislation on pesticides. The new publication will also replace the present MAFF Reference Book 380, *Approved Products for Farmers and Growers*, which currently lists details of all products which have been approved by the Agricultural Chemicals Approval Scheme under chemical names in a very convenient way.

Herbicides

Active ingredient	*Product*	*Supplier*
aminotriazole	Weedazol	Bayer
benazolin + 2,4-DB + MCPA	Legumex Extra	FBC
bentazone + dichlorprop	Basagran DP	BASF
ditto + cyanazine	*ditto* + Fortrol	Shell
benzoylprop-ethyl	Suffix	Shell
bifenox + linuron	Alibi FL	Farm Protection
bifenox + mecoprop	Ceridor	Elanco
bromoxynil + clopyralid	Vulcan	Farm Protection
bromoxynil + dichlorprop	Cornoxynil	FBC
bromoxynil + ioxynil	Stellox 400 EC	Ciba-Geigy
	Hobane	Farm Protection
	Novacorn	Farmers Crop Chemicals
	Deloxil	Hoechst
	Oxytril CM	May & Baker
	Topper 2+2	Union Carbide
bromoxynil + ioxynil + benazolin	Asset	FBC
bromoxynil + ioxynil + dichlorprop	Certrol E	Chafer
	Oxytril P	Murphy
bromoxynil + ioxynil + isoproturon	Doublet	May & Baker
	Twin-Tak	May & Baker
bromoxynil + ioxynil + mecoprop	Brittox	May & Baker
	Swipe 560 SCW	Ciba-Geigy
bromoxynil + ioxynil + trifluralin	Masterspray	Pan Britannica

chlorsulfuron + ioxynil + bromoxynil	Glean TP	Du Pont
chlorsulfuron + methabenzthiazuron	Glean C	Du Pont
chlorsulfuron + metsulfuron-methyl	Finesse	Du Pont
chlortoluron	Dicurane	Ciba-Geigy
clopyralid + bromoxynil	Vulcan	Farm Protection
clopyralid + bromoxynil + mecoprop	Broxolon	Farm Protection
clopyralid + bromoxynil + ioxynil + mecoprop	Crusader	Murphy
clopyralid + ioxynil	Escort	Murphy
clopyralid + ioxynil + mecoprop	Harrier	ICI Plant Protection
cyanazine	Fortrol	Shell
cyanazine + clopyralid	Coupler	Shell
cyanazine + MCPA	Envoy	Shell
cyanazine + mecoprop	Cleaval	Shell
2,4-D	*various*	*many*
2,4-DB	Campbells 2,4-DB Embutox	Campbells Hortichem
2,4-DB + linuron + MCPA	Alistell	Farm Protection
2,4-DB + MCPA	Agrichem DB Plus Redlegor Farmon 2,4-DB Plus Clovacorn Extra Embutox Plus	Agrichem Campbell Farm Protection Farmers Crop Chemicals May & Baker
2,4-DB + mecoprop	Bellco Nintex	Campbell Shell
dicamba + mecoprop	Hyban Di-Farmon M MSS Mircam	Agrichem Farm Protection Mirfield
dicamba + mecoprop + MCPA	Springcorn Extra Banlene Plus Field Marshall Herrisol MSS Mircam Plus Tetralex Plus	Farmers Crop Chemicals FBC Campbell Bayer Mirfield Shell
dichlorprop	Redipon Farmon 2,4-DP MSS 2,4-DP	Campbell Farm Protection Mirfield
dichlorprop + bromoxynil	Cornoxynil	FBC
dichlorprop + 2,4-D	Polymone X	Universal Crop Protection
dichlorprop + MCPA	BASF DP/MCPA Amine 50 Redipon Extra Cleanacres 2,4-DP Plus Farmon 2,4-DP + MCPA Seritox 50 MSS 2,4-DP + MCPA	BASF Campbell Cleanacres Farm Protection May & Baker Mirfield

dichlorprop + MCPA + clopyralid	Mayclene Lontrel Plus	FBC ICI Plant Protection
dichlorprop + MCPA + dicamba	Mephetol Extra	Chafer
dichlorprop + MCPA + ioxynil	Certrol PA	Chafer
dichlorprop + MCPA + 2,3,6-TBA	Camtox	Campbell
dichlorprop + mecoprop	New Formula CMPP/DP Campbells CMPP/DP Ester Clenecorn Plus	Campbell Campbell Farmers Crop Chemicals
dichlorprop + mecoprop + 2,3,6-TBA	Wintacorn Extra	Farmers Crop Chemicals
diclofop-methyl	Hoegrass	Hoechst
difenzoquat	Avenge 630	Cyanamid
dinoseb amine	*various*	*many*
dinoseb-acetate	Ivosit	Hoechst
flamprop-M-isopropyl	Commando	Shell
flamprop-methyl	Lancer	ICI Plant Protection
fluroxypyr	Starane 2	Murphy
fluroxypyr + ioxynil	Stexal	May & Baker
fluroxypyr + ioxynil + bromoxynil	Advance	ICI Plant Protection
glyphosate	Roundup Sting	Monsanto Monsanto
ioxynil	Mate Topper	May & Baker Union Carbide
isoproturon	Hytane 500L Arelon Liquid Tolkan Liquid	Ciba-Geigy Hoechst May & Baker
isoproturon + dichlorprop + bentazone	BAS 462	BASF
isoproturon + ioxynil + bromoxynil	Twin-Tak	May & Baker
isoproturon + ioxynil + bromoxynil + mecoprop	Dictator	Ciba-Geigty
isoproturon + ioxynil + mecoprop	Post-Kite Musketeer Assasin	FBC Hoechst ICI Plant Protection
isoproturon + tri-allate	Trigger	May & Baker
isoproturon + trifluralin	Pre-Kite	FBC
isoxaben	Flexidor	Elanco
linuron	*various*	*many*
linuron + bifenox	Alibi	Farm Protection
MCPA	*various*	*many*
MCPB	Bellmac Straight Chafers MCPB Tropotox	Campbell Chafer May & Baker
MCPB + MCPA	Bellmac Plus Farmon MCPB Plus	Campbell Farm Protection

	Tropotox Plus Trifolex-Tra	May & Baker Shell
mecoprop	*various*	*many*
mecoprop + clopyralid	Seloxone	ICI Plant Protection
mecoprop + ioxynil (esters)	Mylone	Union Carbide
mecoprop + ioxynil + benazolin	Springclene 2	FBC
mecoprop + MCPA + 2,3,6-TBA + dicamba	Cambilene	FBC
methabenzthiazuron	Tribunil	Bayer
metoxuron	Dosaflo	Farm Protection IC Plant Protection Sandoz
metoxuron + simazine	Hermes	Atlas
metsulfuron-methyl	Ally	Du Pont
paraquat	Scythe Grammoxone 100	Cyanamid FBC and ICI Plant Protection
paraquat/diquat	Cleansweep	Shell
pendimethalin	Stomp 330	Cyanamid
terbutryne	Prebane 500L	Ciba-Geigy
tri-allate	Avadex BW Avadex BW Granular	Monsanto Monsanto
trifluralin + linuron	Janus Trifluron Twinpack Elanco Tri-Farmon Marksman Warrior Linnet	Atlas Campbell Chandor Farm Protection Farmers Crop Chemicals ICI Plant Protection Pan Britannica
trifluralin + terbutryne	Terbalin Tripartaurel	BASF Tripart FC
trifluralin + trietazine + linuron	Pre-empt	FBC

Fungicides

Active ingredient	*Product*	*Supplier*
benodanil	Calirus	BASF
benomyl	Benlate	Du Pont
captafol	*various*	*many*
carbendazim	*various*	*many*
carbendazim + 'activator'	Supercarb	Pan Britannica
carbendazim + chlorothalonil	Bravocarb	SDS Biotech
carbendazim + dithiocarbamate	Delsene M Kombat Septal Multi W	Du Pont Hoechst FBC Pan Britannica
carbendazim + maneb + sulphur	Bolda	Farm Protection
chlorothalonil	Bravo 500 Bombadier	BASF and SDS Biotech Universal Crop Prot.

	Tripart Faber	Tripart Farm Chemicals
ethirimol	Milstem	ICI Plant Protection
ethirimol + captafol	Junospore	Atlas
fenpropimorph	Corbel Mistral	BASF May & Baker
fenpropimorph + iprodione	Sirocco	May & Baker
fenpropimorph + carbendazim	Early Impact	ICI Plant Protection
flutriafol + captafol	Impact Extra	ICI Plant Protection
flutriafol + ethirimol + thiabendazol	Ferrax	ICI Plant Protection
iprodione	Rovral Flo	May & Baker
manganese + zinc dithiocarbamate	Trithac FBC Protectant Fungicide	Murphy FBC
nuarimol	Triminol	Elanco
prochloraz	Sportak	FBC
prochloraz + carbendazim	Sportak Alpha	FBC
propiconazole	Radar Tilt	ICI Plant Protection Ciba-Geigy
propiconazole + carbendazim	Hispor	Ciba-Geigy
propiconazole + tridemorph	Tilt turbo	Ciba-Geigy
propineb	Antracol	Bayer
pyrazophos	Missile	Hoechst
sulphur	*various*	*many*
thiophanate-methyl	Cerobin Liquid	May & Baker
triadimefon	Bayleton	Bayer
triadimefon + captafol	Bayleton CF	Bayer
triadimefon + carbendazim	Bayleton BM	Bayer
triadimenol	Bayfidan	Bayer
triadimenol + carbendazim	Bayfidan BM	Bayer
triadimenol + fuberidazole	Baytan	Bayer
triadimenol + fuberidazole + imazalil	Baytan IM	Bayer
tridemorph	Bardew Calixin Ringer	FBC BASF Pan Britannica
tridemorph + carbendazim + maneb	Cosmic	BASF
triforine	Triforine LST Saprol	Shell Shell

Insecticides

Active Ingredient	*Product*	*Supplier*
carbophenothion	Mist-O-Matic Trithion Liquid Seed Treatment	Murphy

chlorfenvinfos	Birlane Liquid Seed Treatment	Shell and ICI Plant Protection
	Birlane 24	Shell
	Sapecron	Ciba-Geigy
chlorpyrifos	Dursban 5G (granules)	Dow and Farm Protection
	Dursban 4	Dow and Murphy
	Spannit	Pan Britannica
deltamethrin	Decis	Hoechst
demeton-S-methyl	*various*	*many*
dimethoate	*various*	*many*
fenitrothion	Dicofen	Pan Britannica
fenvalerate	Sumicidin	Shell
fonofos	Dyfonate MS	Stauffer
	dyfonate IOG (granules)	Stauffer
gamma-HCH	Gamma-Col	ICI Plant Protection
	Gammasan 30 (seed treatment)	ICI Plant Protection
	P.P Mini Leatherjacket Pellets	ICI Plant Protection
gamma-HCH + organomercury	Lindacol	Shell
	Mergamma (seed treatment)	ICI Plant Protection
	Mist-O-Matic Wireworm Liquid Seed Treatment	Murphy
	New Kotol	Shell
heptenophos	Hostaquick	Hoechst
metaldehyde	*various*	*many*
methiocarb	Draza	Bayer
omethoate	Folimat	Bayer
oxydemeton-methyl	Metasystox R	Bayer
permethrin	Decis	Hoechst
phosalone	Zolone Liquid	May & Baker
pirimicarb	Aphox	ICI Plant Protection
	Atlas Pronto	Atlas
	FBC Pirimicarb 50	FBC
pirimiphos-methyl	Blex	ICI Midox
quinalphos	Savall	Sandoz
thiometon	Ekatin	Farm Protection and Sandoz
triazophos	Hostathion	Hoechst
Growth regulators		
chlormequat	*various*	*many*
2-chloroethylphosphonic acid (ethephon)	Cerone	ICI Plant Protection, Marks and Union Carbide
mepiquat chloride + 2-chloroethyl phosphonic acid	Terpal	BASF

INDEX